Thermoinfocomplexity

A Comprehensive Theory of Origin of Life and Complex Adaptive Systems

2020 Edition

Behzad Mohit, M.D.

© 2020

Behzad Mohit, 1933 -

Thermoinfocomplexity

A Comprehensive Theory of

Origin of Life and Complex Adaptive Systems

2020 Edition

Includes bibliographical references.

Revised Edition 2020

Edited by Robert Kain

ISBN-9798692456076

Library of Congress: 2015906716

Cover art by Payam Jafari. payamjafari20@gmail.com

Contents

To the memory of my mother, Sedi Shahroodi,
who always let me ask why and showed me the way
to love and serve people.

Acknowledgements

Writing this book has been an endeavor of exploration, learning, and discovery. It began in 2000, with my curiosity about the physiological, historical, and neurochemical evolution of male aggression and war. My assistants, Sally Wilson and Stephen Roach, helped me conduct an extensive survey into the literature related to this topic. I wrote extensively on this subject, though never completed a publication. I would like to thank them deeply, for their help led me to an inquiry into the appropriate thesis of this book.

From there, I shifted my focus towards an exploration into the evolutionary and neurophysiological basis of aggression vs. altruism and love. The study of altruism, and its evolution, drove my curiosity, leading me to evolutionary biology, where I set out to determine the physicochemical basis of cooperative behaviors. This search expanded to the exploration of evolutionary biology in general. In 2005, I was assisted by Drew Halley, followed by Evan Winchester, both of whom I am indebted to for their extensive research and writing along those lines. Drew went on to pursue a PhD in evolutionary biology and Evan remained to assist me in the ongoing process of research and writing, until the present book's completion.

As the direction of discovery led me towards the more foundational sciences, Anna Ankirskaia, a young and brilliant graduate in biochemistry, helped me research the chemical foundation of life. I owe a great deal to her tireless work. Simultaneously, my other assistants, Dr. Arnas Palaima and Dr. Steven Edmond Kelley helped me incorporate the fields of ecology, as well as recent developments in evolutionary biology. A deeply appreciated contribution, their work provided a sound descriptive foundation for the story of life's emergence. At this juncture, I began to delve into the statistical mechanics, physics, and mathematics underlying the evolutionary emergence of complex adaptive systems, in general, and living organisms, in particular. Jeremy Wheeler carried out a brilliant exploration into the relevant literature and helped me construct my argument. I am sincerely grateful to his participation. Here, I am once again obliged to thank Evan, who in addition to researching the relevant literature, helped edit my disparate writing. Finally, the completion of this new and revised edition would not have been possible without detailed discussions, editorial input, and the enthusiastic encouragement of Robert Kain, my friend. The completion of this work would not have been possible without the extraordinary editorial assistance of Trevor Cohen who joined me in the final stretches of this book.

In closing, I must also acknowledge my profound gratitude for Prof. John Scales Avery, whose 2003 book on information theory and evolution, proved to be a turning point in my thinking on the physicochemical basis and emergence of

2

complex adaptive systems. I am immensely indebted to him for his careful reading of the 2011 version of this book, along with his very kind and encouraging feedback. His flattering review and useful input, on the final 2013 version, proved invaluable at the end of this journey. His encouragement and friendship is most sincerely acknowledged.

Preface

Claude Shannon (1916-2001) is usually considered to be the "father of information theory." In 1949, motivated by the need of AT&T to quantify the amount of information that could be transmitted over a given line, Shannon published a pioneering study of information as applied to communication and computers.

When Shannon had been working on his equations for some time, he happened to visit the mathematician John von Neumann, who asked him how he was getting on with his theory of missing information. Shannon replied that the theory was in excellent shape, except that he needed a good name for "missing information". "Why don't you call it entropy?" von Neumann suggested. "In the first place, a mathematical development very much like yours already exists in Boltzmann's statistical mechanics, and in the second place, no one understands entropy very well, so in any discussion you will be in a position of advantage!"

Shannon took von Neumann's half-joking advice and used the word "entropy" in his pioneering paper on information theory. Missing information in general cases has come to be known as "Shannon entropy." But Shannon's ideas can also be applied to thermodynamics:

Consider an ensemble of N weakly-interacting identical subsystems. For simplicity, we might imagine that there are only two possible quantum states for each subsystem, 1 and 2. If we know how many subsystems are in state 1 and how many in state 2, then we know the macrostate of the ensemble. If we know precisely which of the subsystems is in each state, then we know the microstate. In a given macrostate, the missing information that would be needed to specify the microstate can be shown to be proportional to the thermodynamic entropy of the ensemble. To give it a name, we can call it "missing thermodynamic information." Its negative, we can call "thermodynamic information."

From the discussion given above we can see that there is a close relationship between cybernetic information and thermodynamic information. However, despite the close relationships, there are important differences between Shannon's quantities and those of thermodynamics and statistical mechanics. Cybernetic information (also called semiotic information) is an abstract quantity related to messages, regardless of the physical form through which the messages are expressed, whether it is through electrical impulses, words written on paper, or sequences of amino acids. Thermodynamic information, by contrast, is a temperature-dependent and size-dependent physical quantity. Doubling the size of the system changes its thermodynamic information content; but neither doubling the size of a message written on paper, nor warming the message will change its cybernetic information content.

Furthermore, many exact copies of a message do not contain more cybernetic information than the original message.

The evolutionary process consists in making many copies of a molecule or a larger system. The multiple copies then undergo random mutations; and after further copying, natural selection preserves those mutations that are favorable, as emergent complexity. It is thermodynamic information that drives the copying process, while the selected favorable mutations may be said to contain cybernetic information. The cybernetic information distilled in this process is always smaller than the quantity of thermodynamic information required to create it (both measured in bits) since all information must have a source.

One of the most important innovations in the application of thermodynamics to chemistry was the definition of a quantity which we now call "Gibbs free energy." This quantity allows one to determine whether or not a chemical reaction will take place spontaneously. Chemical reactions usually take place at constant pressure and constant temperature. If a reaction produces a gas as one of its products, the gas must push against the pressure of the earth's atmosphere to make a place for itself. In order to take into account the work done against external pressure in energy relationships, the German physiologist and physicist Hermann von Helmholtz (1821-1894) introduced a quantity (which we now call heat content or enthalpy) defined by $H = U + PV$ where U is the internal energy of a system, P is the pressure, and V is the system's volume. The American scientist John Willard Gibbs (1839-1903) went one step further than Helmholtz, and defined a quantity which would also take into account the fact that when a chemical reaction takes place at constant temperature, heat is exchanged with the surroundings. Gibbs defined his free energy by the relation $G = U + PV - TS$ where T is the absolute temperature and S is the entropy. Gibbs was able to show that the second law of thermodynamics, which says that the entropy of the universe always increases, implies that for a spontaneous chemical reaction taking place at constant temperature and pressure, the Gibbs free energy of the reacting system always decreases.

The Second Law of Thermodynamics, which states that the entropy (disorder) of the universe always increases, seemingly contradicts the high degree of order and complexity produced by living organisms. This apparent contradiction has its resolution in the information content of the Gibbs free energy which is constantly entering the biosphere from outside sources. The earth's biosphere is not a closed system. It is constantly receiving thermodynamic information in the form of photons from the sun.

When a photon from the sun reaches (for example) a drop of water on the earth, the initial entropy of the system consisting of the photon plus the drop of water is smaller than at a later stage, when the photon's energy has been absorbed and shared among the water molecules, with a resulting very slight increase in the temperature of the water. This entropy difference can be interpreted as the quantity of thermodynamic information which was initially contained in the photon-drop system, but which was lost when the photon's free energy was degraded into heat.

For example, if the photon energy is 2 electron-volts, and if the water drop is at a temperature of 298.15 degrees Kelvin, then ΔS = 112.31 bits; and this amount of thermodynamic information is available in the initial state of the system. In our example, the information is lost; but if the photon had instead reached the leaf of a plant, part of its energy, instead of being immediately degraded, might have been stabilized in the form of high-energy chemical bonds.

In his important book *What is Life?*, the Austrian physicist Erwin Schrödinger (1887-1961) says: "What is that precious something contained in our food which keeps us from death? That is easily answered. Every process, event, happening, call it what you will; in a word, everything that is going on in Nature means an increase of the entropy of the part of the world where it is going on. Thus, a living organism continually increases its entropy, or, as you may say, produces positive entropy, which is death. It can only keep aloof from it, i.e., alive, by continually drawing from its environment negative entropy, which is something very positive as we shall immediately see. What an organism feeds upon is negative entropy. Or, to put it less paradoxically, the essential thing in metabolism is that the organism succeeds in freeing itself from all the entropy it cannot help producing while alive..."

"Entropy, taken with a negative sign, is itself a measure of order. Thus, the device by which an organism maintains itself stationary at a fairly high level of orderliness (= fairly low level of entropy) really consists in continually sucking orderliness from its environment."

Schrödinger's deep insights give us a picture of the source of the seemingly miraculous complexity and order that we witness around us: A flood of information-containing free energy reaches the earth's biosphere in the form of sunlight. Passing through the metabolic pathways of living organisms, this information keeps the organisms far away from thermodynamic equilibrium ("which is death"). As the thermodynamic information flows through the biosphere, much of it is degraded into heat, but part is converted into cybernetic information and preserved in the intricate structures which are characteristic of life. The principle of natural selection ensures that as this happens, the configurations of matter in living organisms constantly increase in complexity. This is the process which we call evolution or, in the case of human society, progress.

I hope that these remarks will serve as an introduction to Dr. Behzad Mohit's excellent book that brings into light the combined role of thermodynamics, information and complexity theories in explaining the emergence of life and evolution of complex adaptive systems.

John Scales Avery

October 2013

Author's note on the extension and expansion of the concept of information described in the above Preface:

Here I would like to introduce two concepts. Given that information, as defined above by Avery, is measured at room temperature: (1) we would like to extend that relationship of information and energy over a range of temperatures, and (2) we theorize that information carries both energy and mass, and we name this quantum measure of information the *infon*. This second concept is described in the following.

It is a broad and deep departure from any concept of information available in the literature up to this date. The details of this definition and conjecture will be published under a separate cover under the title of "The Natural Quantum Theory of Information."

Conjecture About the Structure of Information

An infon can be defined by the following characteristics:

1) The infon is a quantum particle (particle-wave duality) with mass and kinetic energy.

2) The infon has an internal structure that carries information.

3) The attractive force between infons is the weakest in the Universe.

4) The infon interacts weakly with all other elementary particles, and exchanges energy and information.

5) The infon is the building block of all other elementary particles.

Implications of the infon:

1) Energy and information carried by the infon drive the evolution of the Universe.

2) Infonic Field Theory (to be published under a separate cover) may subsume Quantum Field Theory, and replace General Relativity altogether, potentially providing explanations for dark matter, dark energy, black holes, etc.

Introduction

The eternal mystery of the world is its comprehen-
sibility...The fact that it is comprehensible is a miracle.
-Albert Einstein

Comprehension is an emergent phenomenon of energy
flow through a neurochemical network.
-Behzad Mohit

Give me a place to stand, and I will move the earth.
-Archimedes

The question of how life emerged on Earth is one that can sustain a lifetime of inquiry. It was this question that motivated Darwin, upon returning from the Galapagos, to revisit his study and spend the next twenty years writing *On the Origin of Species*. Since Darwin, the theory of biological evolution has been presented from many different perspectives, from the writings of Erwin Schrodinger and Stephen Jay Gould, to John Maynard Smith. The scope of evolutionary theory, brilliantly articulated by these eminent thinkers, is an impressive edifice, expanded throughout the 20th century to encompass the entirety of the biological domain, from the molecular level to that of species. Now we expand it from its cosmic origins to global ecology.

Despite the thoroughness with which Darwin's ideas have been integrated into biology, evolutionary theory continues to only describe the existence of life, in terms of the persistence of life. It essentially restates a premise that has remained largely unexamined since Darwin first posited it—that life exists in its present forms because these forms succeeded in propagating against the headwinds of natural selection. Extinction is the product of a failure to do so. The extraordinary diversity of organisms arose from repeated iterations of this process. Biological units, from microbes to mammals, seek to replicate themselves into the next generation, struggling against the odds of survival in a hostile world of competitors. This circular argument—that evolutionary fitness is the reason for survival, and survival is the measure of fitness—is the core mechanism of evolution by natural selection. Improbably structured organisms emerge against the indifferent, yet menacing backdrop of nature.

The mechanisms of evolutionary change that rely on this circular logic, many of which constitute the NeoDarwinian synthesis, are incomplete. They describe *what* has occurred over the course of evolutionary history, but they seem to miss the *how*. Specifically, they fail to explain how complex life emerged from the atomic and molecular levels of organization. In short, these proximate explanations provide a necessary, but insufficient set of principles with which to describe the emergence of life. More to the point, natural selection, defined as the differential survival rates propagated through the generations, is more an observed fact than an explanation of the physicochemical mechanisms operative in the process of natural selection. Today, evolutionary theory remains to a large extent a descriptive means by which to account for the origins of life. At a more comprehensive level, there has been little agreement with respect to satisfactory physiochemical processes underlying the evolutionary process. Biologists have tended to shy away from positing a physical basis for the process of evolution. As such, the "how" of evolution is a question that has been, in this author's view, inadequately explored.

In recent years, however, several works have sought to bring the discussion of biological evolution beyond the neoDarwinian synthesis by integrating the concepts of energy and thermodynamics into evolutionary theory. John Avery has elucidated the importance of the convertibility of energy and information in the evolutionary process.[1] Schneider and Sagan have summarized research at the intersection of thermodynamics and biology, revealing how the processes of biological evolution are intimately related to the emergence of flow systems or dissipative structures, which seek to maximize entropy production in molecular systems. These systems tend to avoid thermodynamic equilibrium.[2] Ricard Solé and Brian Goodwin have given an overview of the non-linear dynamics that characterize biological systems, describing how the dynamics of complexity pervade biology.[3] However, until this writing, there has been no comprehensive synthetic theory of evolution to adequately address the question of "how," in physicochemical

15

and mathematical terms, life's complex adaptive systems have evolved.

Now, in the autumn of my life and after extensive research in many scientific fields, I believe that I have reached a "eureka" moment. This book is my humble attempt to share the insights I have gained over the course of a lifetime's inquiry into the "magic" of life's emergence across all strata of existence, from the first singularity, to the big bang, then to molecules and modern societies. This book presents a comprehensive synthesis of developments that apply the laws of probability, thermodynamics, complexity, and information theories to the question of evolution. In the course of my work, I coined the term Thermoinfocomplexity to designate this new synthesis, radically extending evolutionary theory to encompass both quanta and quasar, in the process of evolution by stochastic selection.

On the Shoulders of Giants

The theoretical foundations of this work have been developed gradually over the past couple of centuries by scientists in the fields of thermodynamics, evolutionary theory, statistical mechanics, information theory, and complexity theory. The first two, thermodynamics and evolutionary theory, emerged in the wake of the Industrial Revolution of the late 18th and early 19th centuries. This period was perhaps the most significant turning point in history, in terms of human impact

16

upon the biological, population, and the economic dynamics of life on Earth.

Thermodynamics, the study of the flow of energy through physical systems, came about through the desire to improve the efficiency of steam engines. The field of thermodynamics has evolved in parallel with our ability to create increasingly efficient machines, but it has applications that extend well beyond the study of engineering. Darwin published the seminal works that laid the foundation for evolution by natural selection around the same time that scientists worked out the laws of classical thermodynamics. We shall see that these two principles, natural selection and thermodynamics, although intimately related, never crossed paths in their early history.

Later it became clear that thermodynamic principles as universal laws, must be incorporated into any comprehensive scientific theory, including that of evolution. The First Law of Thermodynamics states that the total energy of any system is conserved. It can change form, but the overall energy and matter content of a system can neither be created nor destroyed. The Second Law of Thermodynamics, introduced by Rudolf Clausius, states that the overall direction of energy flow in any system must be in the direction of increasing entropy, a term that has traditionally been defined as a measure of the "disorder" of a system. Clausius's observation that heat always flows from warm to cold and never the other way around, gave a universal direction to time's arrow. The universe, and by extension life, is an irreversible process, always tending toward increasing entropy. We will see that life forms emerge through their ability to find a niche between the warmth of their internal structures and the relative cold of their surroundings, creating systems that temporarily slow this flow of increasing entropy.

Unbeknownst to Darwin, Gregor Mendel contemporaneously developed a genetic theory of inheritance, the necessary complement to Darwinian selection. Mendel's work languished in obscurity until it was rediscovered around the turn of the 20th century, providing the "inheritance factors" that later came to be known as genes. Equipped with the knowledge of genetics, evolutionary biologists spent the remainder of the 20th century elucidating the molecular basis of evolution. Watson, Crick, and Franklin's discovery of DNA in 1953 provided a molecular model of the mechanism of inheritance. DNA formed the foundation for genetics, and the discovery of the genetic code revealed DNA's capacity to bind information. The code itself, rather than any object or molecule, emerged as the primary unit of genetic inheritance. This enabled us to understand living matter as having an inherent capacity to encode information within it.

The desire to understand the genetic code, and by extension life itself, is what propelled the massive investment of time and resources into the human genome sequencing project as well as other, similar inquiries into the nature of the genetic code. Comparative sequence analysis of the genomes of various species has allowed scientists to reconstruct a detailed picture of the evolutionary tree of life on a molecular basis. Nevertheless, their findings only describe *what* happened over the course of evolution, rather than *how* it came to be. The question of *how* remains a difficult one. We are coming to discover that the prediction of protein and biological structures from the DNA sequence alone is itself challenging, due to the massive complexity of the ensemble of molecules making up the living matter. The emergence of complex living organisms depends upon the incorporation of information into the flow of thermodynamic energy. By incorporating thermodynamics and information theory into the study of questions of the emergence of life, I will show how prebiotic atoms and molecules evolved into complex life forms. I will also show that these complex forms emerge from their constituent parts following the most statistically probable course.

From the mid-20th century to the present, information theory has developed a rigorous mathematical understanding of communication, providing a quantitative basis for the study of information transfer within and between systems. Somewhat surprisingly, we believe information theory is intimately related to the concept of thermodynamic entropy. The intersection of these two fields of inquiry is one of the most exciting areas of contemporary scientific research. Both theories are grounded in the mathematical methods of probability, which have left their mark on just about every area of scientific study. And in this edition, we expand upon the mathematical basis that underlies both of them.

The application of probability theory to the study of biological organisms and their evolution, has been the basis of much research into life's origins, but as with the application of molecular approaches to biology, much of this work has been of a descriptive nature. However, when information theory and thermodynamics are integrated to explain the problems of the emergence, evolution, and maintenance of life, we gain a profound new perspective. I hypothesize, following the work of John Avery and Annila, that this integration provides a basis for the equivalence of information and energy. Information contains energy and energy contains information. Complex systems, from atoms to networks of organisms, capture this energy in information networks that sustain the ordered complexity that appears in the organization of all life.

The emergent theories of chaos and complexity are the foundation of the study of life and other complex adaptive systems. In the past, the deterministic theories and predictive equations of physics and chemistry, the so-called "hard sciences," have always remained beyond the purview of biological theory. In this edition of the book, problems in biology can be studied in depth through the seamless connection between cosmological quanta and the most probable complexity that also constitutes living matter.

The mathematician and meteorologist Edward Lorenz developed the foundations of chaos theory by applying computational approaches to the study of weather prediction. The problems of meteorology are explained by the theories of chaos and complexity. Similarly to weather prediction, in the emergence of complex systems such as living matter, small changes in the initial conditions result in wildly divergent outcomes.

The solutions to these problems can be determined by applying probability and thermodynamic theories to patterns known as "strange attractors." Up until this writing, the problems of biology have not been explained by theories of physics. Though unsolvable in the deterministic sense, we will show how strange attractors illustrate the path of emergence from its most abstract mathematical origin (i.e., asymmetry). Like hurricanes and cyclones, life is a system that comes into existence, has an identifiable form, and dissipates into the atmosphere from which it emerged. We will see how pathways of information and energy flow integrate through positive and negative feedback loops in the structure and behavior of complex adaptive systems.

A New Vision

All of these perspectives converge upon the question of evolution. Taken together, the principles of thermodynamics, the new quantum information theory, and complexity theory combine to give us a new theory of life, as emergence: the new Thermoinfocomplexity theory. To avoid the confusion generated by attempting to distinguish between living and nonliving systems in the traditional manner, the phrase "complex adaptive systems" will be used throughout this book to describe the coherent structures that emerge as the adaptive response to the environment at various levels of complexity. Complex adaptive systems, from eddies in a river to the global economy, emerge through critical points of interaction between matter, information, and energy. They persist so long as there is available energy to maintain the coherence of these entities, and they dissipate once the available energy used to organize the system has been depleted.

Throughout this book, we will extend the meaning of metabolism to include any observed use of available energy by a complex adaptive system to maintain its dissipative structure. Once the available energy is metabolized by the system, it can either collapse into the ground from which it emerged, or couple with another system in a symbiotic response to the depletion of energy in its environment. These complex systems are additive. If there is an additional available source of energy that can be accessed by the coupling of two or more systems, entrained cooperation can emerge between systems, resulting in a higher order of complexity. The increased energy efficiency, formed in response to the depletion of an existing energy source, is the product of the resonant coupling of periodic oscillations in the metabolic pathways already established by the previously independent systems. The entrainment of these oscillations manifests in strange attractor pathways that reflect the newly coordinated metabolic activity of the emergent system. This is a process akin to respiration or photosynthesis. It is periodic, but not linear or deterministic in nature.

Within this vision, there is no need to invoke a designer or to recreate a contingent historical narrative in order to explain the emergence of the bountiful and dazzling complexity of life. The electron clouds of atoms, the alternating bands in seashells, the pulsations of human hearts, and the circadian rhythms of fruit flies, stripped of the artificial divisions between them, come to make sense. Each occupies its respective position along the energy gradient between the sun and the earth. Along that path, we encounter innumerable discrete forms of complex adaptive systems emerging, dissipating, or merging to become more complex. This process is evolution.

We note that each system is an integrated network of information within matter. It persists in time and space through a continuous flow of Gibbs free energy. We further note throughout this book, that in nature, energy scarcity drives the emergence of more complex entrained systems. It is indeed the complexity and redundancy of the entrained information networks of a complex system that gives it its selective advantage to persist in time. In this context, Thermoinfocomplexity is a descriptive term, depicting the stochastic emergence of complex adaptive systems in nature. It is a comprehensive theory of *how*, throughout astronomic timescales, chance encounters of energy and matter and their interconversions have progressed from sun to earth, from photon to Gaia, in a process that includes us as conscious awestruck observers.

In the following chapters, I will introduce several observations explained by the principles of Thermoinfo-complexity. As the book progresses, so does the complexity of the evolutionary system discussed. Starting with its mathematical basis, I go on to discuss the emergence of life, multi-cellularity, social networks, civilizations, and superorganisms. The final chapter applies these observations on the emergence of complexity to speculate on its future course.

Chapter 1 introduces some of the major concepts of Thermoinfocomplexity theory, such as energy efficiency, natural information, complexity, and emergence. It describes how complex adaptive systems become more complex, operate at a maximum of energy efficiency, and hence are selected to persist over time when energy is scarce. A Natural Theory of Information is hypothesized which replaces Shannon's classical notion of information. A new Quantum Theory of Complexity is developed which gives precise definition to the ideas of complexity, emergence, and complex adaptive systems. As a separate writing, we will formally define information and show its connection with energy.

Chapter 2 develops the hypothesis of a Quantum Theory of Information that is consistent with the notion that a system with more natural information is one with lower entropy and higher internal structure. We show the connections between energy and information, which will be the foundation for emergence of complex adaptive systems. An information manifold is created that serves as a network, or underlying structure, of particulate information (Infons), providing emergent structures that follow the efficient Hamiltonian pathways..

Chapter 3 formulates a full Theory of Complexity which had previously eluded scientists. Precise terms are developed which capture the common intuitions we have about complexity, emergence, and complex adaptive systems. Using this new Theory of Complexity, the laws of Thernoinfocomplexity and a 'New Theory of Everything' can be formulated.

Chapter 4 introduces the theorems of Thermoinfo-complexity, including a theorem that states that complexity must increase in systems where there is a suitable energy source, such as the energy gradient induced by solar radiation on Earth. Other theorems are discussed which explain how species persist and how punctuated equilibrium arises in evolution. Fundamental principles of Thermoinfocomplexity are summarized in this chapter.

Chapter 5 explains the mechanisms of how complex adaptive systems evolve through fluctuations in Gibbs free energy available in the environment. It goes on to discuss the roles of catalysts, autocatalytic structures, and attractors in evolution.

Chapter 6 deals with the origin of complex adaptive systems and describes the physical basis of the evolutionary selection process, beginning with inorganic molecules and detailing their transformation into organic compounds. This chapter describes the emergence of increasingly complex adaptive systems, leading to primitive organisms, such as bacteria.

Chapter 7 explores the principles governing how biological networks, composed of bacteria, biofilms, and the multicellular form of slime mold, emerge in response to the scarcity of Gibbs free energy in the environment. It describes the foundation of elegant biological structures, based on the principles of Thermoinfocomplexity, as outlined in Chapter 4.

Chapter 8 describes the rise of multicellular organisms of greater complexity, cell-to-cell communication, and the process of ontology that mimics evolution, as commonly described (ontogeny recapitulates phylogeny). This chapter further describes examples and consequences of broken systems of biological communication. It develops the correlation between lower metabolic rates and the increased complexity of complex adaptive systems.

Chapter 9 deals with the elegant patterns of cooperation present in complex animal behaviors, from flocks of birds and schools of fish to packs of wolves. It demonstrates how energy-efficiency underlies evolved group behaviors.

Chapters 10 and 11 describe the physical-chemical basis of cooperation through communication, from ants to bats, whales, dolphins, and humans. They elucidate the mechanism of cooperation between organisms, and its relation to the role of memory and chemical signaling. These chapters clearly explain the process of cooperation, altruism, and bonding between organisms.

Chapter 12 provides a detailed description of how energy flows through the human superorganism, from hunter gatherer tribes to cities and states. Furthermore, the hierarchical structure of human superorganisms is generalized and described through the lens of Thermoinfocomplexity. It is an attempt to describe and understand the complexities of human social organization, as a response to ebbs and flows in the environmental availability of Gibbs free energy.

Finally, Chapter 13, "Imagine" is a vision of the emergent global network of the human superorganism. It predicts the formation of an efficiently networked amalgamation of man and machine, leading to a happy and intimately integrated, global human society.

Chapter 1 –
Thermoinfocomplexity
Theory

Figure 1.1 Lightning bolt showing how the path of least action attracts energy.

1.1 Energy Efficiency

Our earth is full of complex organisms and the level of complexity has increased dramatically over time, from single cell organisms to multi cell organisms, animals and finally humans. Our Universe has also evolved to a highly complex level of organization since the big bang as quantum became atoms, then molecules, and star systems organized into galaxies, clusters of galaxies, and even super clusters. We have discussed in the introduction that Darwin's theory of evolution did not describe the "how" of evolution. This book tells the "how" of evolution of complex matter from the big bang to the emergence of life on Earth.

It might seem as though the Second Law of Thermodynamics creates a difficulty for theories to explain the rise of life and complex organisms. The second law states that entropy (dilution of information) must always increase in a system over time. The corollary of this law is that Gibbs free energy, or the amount of energy available to do work in a system, is decreasing. An increase in entropy is a decrease in complexity, so the Universe and all that is in it is moving towards a global state of lower complexity. If so, how then do complex systems arise?

The second law has often been described as an increase in disorder, an imprecise term which has led to some confusion in understanding the law. A better description follows Boltzmann who formulated the second law in terms of probability and microstates to assert that "systems move towards more probable states." This means the microstates of mixupedness are more probable than states of even distribution of energy. In brief, the second law is saying that systems as a whole move towards mixupedness, since there are more ways to be spread out than there are ways to remain concentrated or evenly distributed. Examples of the law are the movement of heat from warm bodies to colder bodies, or two gases placed into the same container spontaneously mixing, to come to even distribution of their molecules.

The second law, however, does not explain how complex adaptive systems arise. As entropy increases in general and systems mix up, and become less complex, we note however that complex systems can arise on a local basis, sucking order from their environment, while the increase of total entropy continues, thereby not violating the second law.

So, if it's possible for complexity to arise and not violate the Second Law of Thermodynamics, how do these processes occur?

Systems which are most energy efficient are systems of efficient dissipation of energy. In the ups and downs of energy availability in nature, by statistical probability, those configurations that are most energy efficient will persist. By the law of least action, these energy efficient systems, or maximum dissipators of energy see energy channeled to themselves. In the same way that a lightning bolt seeks the least action path and once finding it, fills that path with flows of energy (see *Figure 1.1*), or river basins channel water flow, energy dissipators attract energy.

Figure 1.1 Lightning bolt showing how the path of least action attracts energy [4]

All systems will encounter periods of time where energy is scarce. During processes of energy scarcity, the maximum energy dissipators can still attract enough energy to persist whereas the systems which are not energy efficient will fail to gain the energy they need to persist. Therefore, due to the principle of least action, energy efficiency systems are selected by the law of least action to persist over time.

1.2 Thermoinfocomplexity

In the rise of complex adaptive systems in the Universe, however, there is much more than just the Second Law of Thermodynamics and the principle of least action. By the laws of Thermoinfocomplexity, which we will explain in detail in Chapter 4, we can prove how the complex adaptive systems become more complex, and operate at a maximum of energy efficiency, as they persist over time.

Let's begin by defining complexity, complex adaptive systems, emergence, and emergent complexity before we state the Theory of Thermoinfocomplexity.

An emergent complex system has been often colloquially taken to be one in which the whole is more than the sum of its parts, such as a human being made up of cells, tissues or organs, or a highly-networked system, such as a flock of birds. We therefore developed a concept of "natural information" that captures the following equivalent ideas:

1) the difference between the information of the whole, and the information of the sum of the parts,
2) the degree of interdependence of the parts,
3) how much the parts predict the whole, and
4) the internal information, or structure, of an object.

More precisely we defined the concept of natural information to be the measure of how much more information the emergent system has than its parts, and we define complexity as the expected "natural information content."

By Complex Adaptive Systems, we mean those that persist in their structure over time (we make all these definitions mathematically rigorous in Chapter 2). Molecules, organisms, or an army of ants are all examples of complex adaptive systems. A flock of birds is an example of a complex adaptive system, but half of the flock or a group of birds in a tree as one system is not.

We will take as our motivating concept the idea of an emergent system as a system that is meaningfully *its own entity*, so that it may, for example, follow a set of rules of its own. So, to characterize an entity, we must capture a system whose parts are interdependent, and which contains exactly those parts that happen to be interdependent, and no others. More generally, complexity, or expected natural information, is a measure of the average statistical interdependence of the components of that system. Therefore the "correct" components of a system are precisely those that increase its complexity. This perspective motivates our next definition. A system will be called emergent if it is at a local maximum of complexity.

With our terms defined, we are now ready to describe Thermoinfocomplexity.

Systems in the Universe undergo random stochastic processes, and in the process complexity may emerge. Some of these systems will become complex adaptive systems, which are then selected by the law of least action to survive as energy efficient systems and persist.

Complex adaptive systems complexify and emerge to higher states of complexity. They continue to rise to higher levels of complexity, each level at a local maximum of energy. The complex adaptive systems decrease entropy locally but

34

increase entropy globally in full compliance with the second law of thermodynamics.

1.3 The Three Papers

God doesn't play dice with the universe.
-Albert Einstein

Dice play God with the universe.
-Behzad Mohit

In 2017, I embarked upon an effort to derive with high mathematical rigor the foundational concepts of Thermoinfocomplexity. Never would I have believed what amazing and groundbreaking results would emerge from this effort. I have written three papers (which will be published as a separate work) that have respectively contributed to the development of the following: a new Theory of Information supplanting Shannon's Theory; a new and precise theory of Complexity, Emergence, and Complex Adaptive Systems, which had not heretofore been satisfactorily developed; and then serendipitously, Infonic Field Theory. Infonic Field Theory connects Quantum Mechanics and General Relativity in a grand unification theory of (nearly) everything.

In developing these new theories, we did not need to create any difficult concepts of extra dimensions of space and time or introduce cumbersome ideas such as multiverses, instead, we have grounded the theories in the basic ideas of probability and information. From those simple and intuitive foundations, concepts have emerged that are able to explain how the universe has evolved from the big bang to present day. Einstein wanted a simpler, unified theory from which complexity would emerge logically, sans weirdness.[5] Finally, such a theory has arrived.

The first paper develops the Natural Quantum Theory of Information which replaces the Shannon's classical notion of information in use since 1948. The paper presents a precise mathematical definition of natural information, which we show to be 100% opposite to Shannon's definition. We have developed the theory of natural information in both the classical setting of statistical mechanics, and the quantum mechanical setting, and then demonstrated a deep and precise connection between energy and information. From there we developed topological and geometric aspects of the theory, namely the information manifold and several associated topological invariants.

In the second paper we have been able to give precise definitions of both complexity and emergence in the context of natural (quantum) information. We developed a theory of complexity, emergence, and complex adaptive systems. We prove that systems of higher complexity have greater mutual dependence; that emergent systems are best described "in their own terms"; that complex adaptive systems are emergent over

36

time; and that emergence is a topological invariant of statistical systems.

The third paper develops a new theory of Quantum Mechanics and General Relativity which is equivalent to the existing theories, except it is applicable at sub-millimeter and galactic distances. This new theory is derived from our new theory of information, our information manifold, and the abstraction proof of the relation between information and energy which are then inserted into the main equations of Statistical Mechanics. Lastly a backward extrapolation yields Infonic field theory which is then utilized in the highly surprising and revealing new theoretical formulations. The new statistical approach to gravity explains several of the known discrepancies in Einstein's theory. Gravitational field theory is replaced with Infonic field theory and gravitons are replaced by infons. The notion of curved spacetime is dispensed with in the new theory and only the Minkowski space of special relativity is used.

Chapter 2 – Natural Quantum Information

2.1 Shannon's Theory of Information

2.2 The Natural Quantum Theory of Information

2.3 Information and Topology

Figure 2.1 John Avery's *Information Theory and Evolution,* 2nd Edition.

2.1 Shannon's Theory of Information

Let me say from the outset in order to understand the content of this communication we need to clarify two fundamental concepts. First, entropy, which here I define in a very clear sense. Entropy is a complete lack of information. Second, information is completely opposite to what has historically been defined by the great Claude Shannon. This is such a fundamental difference that I think I need to take a few words to describe what information was, as defined by Claude Shannon, and how I got to the conclusion of its opposite definition, as will be detailed in this writing. In studying John Avery's book (see *Figure 2.1*) on Information and Evolution, I noticed two things. First, the definition of information there was based on Shannon's definition of information and his definition of entropy had followed Boltzmann's.

Figure 2.1 John Avery's *Information Theory and Evolution,* 2nd Edition.

Claude Shannon had a very strong background in communication and arrived at his definition of information from a message being transferred from the sender to the receiver. Shannon initially debated whether he should call the sending and receiving of the message between the sender and receiver "information" or "uncertainty." That already showed a confusing conflation of two terms that were opposite in meaning. But the final blow came from his famous conversation with the mathematician John Von Neumann. "My greatest concern was what to call it," Shannon said. "I thought of calling it 'Information,' but the word was overly used, so I decided to call it 'Uncertainty.' Then John Von Neumann told me that he had a better idea. 'We should call it entropy, for two reasons: in the first place your uncertainty function has been used in statistical mechanics under that name, so it already has a name. In the second place and more important, nobody knows what Entropy really is, so in a debate you will always have the advantage." Thus, thermodynamic entropy became firmly conflated with information and would remain so for decades. I had to take a long journey through Szilard, Wheeler, and others who confuse language dependent communication and semiotics with the abstract notion of information itself. It has been a long painful journey for me to finally come to the realization that information is not entropy and information is not a message. It is an abstract concept very much tied to energy. In fact, it can be said that it is a manifestation of energy. This realization makes a very clean transition between information, complexity, and emergence, as well as creating a clear understanding of the flow of information from abstract energy to the complex quantum world of the cosmos, as well into the form of Gibbs free energy to the complex adaptive systems exemplified by living matter.

Shannon derived his formula using the following reasoning. Suppose Alice sends a message to Bob informing him that aliens have landed in Paris. (Here we also assume that Alice's messages are always correct!) Bob, presumably, will be extremely surprised. That indicates that Alice's message contains a lot of information. If instead Alice had told Bob that the sun would rise tomorrow, Bob would be unimpressed, and the message would carry essentially no information. Therefore, Shannon reasoned, if the message describes an event certain to occur with probability of 1 then the information content of that message is zero. But the more unlikely the event, the more information would be contained in a message saying that an event did in fact happen. Presumably if a message described the occurrence of an impossible event, with probability equal to zero, then the information content would be infinite. Using this approach, and with some extra mathematical assumptions, Shannon was able to prove that his formula for information was, up to a constant multiple, the only equation that can describe his notion of information.

There was only one problem. Shannon's initial reasoning would turn out to be misguided, and so subtly misguided that it would lead mathematicians and physicists astray for decades. The problem was this. Shannon's work on information was motivated by communications. Because of this, his definition was inherently external. You'll notice that we implicitly use Bob's perception of probability to define the

42

information content of a message being sent to him. If we use Alice's perception of probability, the information content of the message would be entirely different. (In fact, it would always be 0, since Alice would always know the content of the message she was sending to Bob.) And more damningly, if we think about Bob himself as already having information about the world (here I mean "information" colloquially, but in a way we will later make precise), then according to Shannon, the information content of the message is only as good as Bob's ignorance. In general, as Bob gets more knowledgeable about the state of the world, the messages sent to him have less and less information. But physicists always prefer objective, or natural, measures. In the above scenarios, they would be less interested in the information content of the message from Bob's point of view than the information content of the message as a stand-alone entity, or perhaps from the universe's view. And from this view, the more knowledge Bob has about the world, the more information content he (as an entity) should be taken to have. Since Shannon's formula measures Bob's ignorance, it therefore measures lack of information (i.e., entropy) from our perspective.

2.2 The Natural Quantum Theory of Information

We solved Shannon's mistake with our definition of the natural information of a system. Our goal with this concept was to provide a definition that captures the true opposite of entropy.

Therefore, colloquially, a system with more natural information ought to be one with lower entropy. Moreover, higher natural information ought to correspond to higher internal structure. Whereas a gas at equilibrium has no statistical structure beyond its lowest-energy distribution, a human has a great deal of statistical structure. Probabilistically we can think of this in terms of the mutual dependence of its components; or via a more familiar expression, the degree to which the whole is more than the sum of its parts. Since we have established that information is the opposite of entropy, as a first step towards defining information, we defined pre-information as the opposite of Shannon information (Shannon entropy).

Our definition of natural information, then, is the difference of the pre-information of the parts of the system X, and the pre-information of the whole. Our formulation measures the following (equivalent) ideas:

- the difference between the whole and the sum of the parts
- the degree of interdependence of the parts
- how much the parts predict the whole
- the internal information, or structure, of an object

It showcases the fact that Shannon's formula measures not information, but its lack. We have also extended the Natural Theory of Information in the classical mechanics setting to include quantum mechanics.

2.3 Information and Topology

If you look at the landscape of the earth, you see it as a tremendous number of wrinkles and topologies. That is like the structure of the brain. What I have thought about as a topological analogy between the brain and the earth, or any other smaller masses, such as the moon or quantum, we note that the larger a mass, the more complex the topology (greater number of wrinkles). The Earth as a whole has an intricate topology that contains a massive amount of information.

This principle is easily justified. Any physical structure that is fixed in mass and volume, but increasing in information, must therefore increase the complexity of that fixed-mass structure. In the simplest example, a smooth structure, such as an infant's brain or an undeveloped planet, must become more wrinkled with time. More generally, any physical system will become more fractal with time, by which we mean not exact self-similarity, but rather possessing detailed structure at more and more levels of resolution (indeed, some authors take this to be an alternate definition of complexity, although it is not one that we find has much mathematical traction). Thus, we are justified in considering, more abstractly, the topology of information. To this end, we have developed an information metric.

Theorem 1:

Given any probability measure, the natural metric induces a smooth manifold structure on the space of objects. If the probability measure is classical and discrete, the manifold is 0-dimensional; but if the measure is continuous or quantum, the manifold is higher-dimensional, with the number of dimensions given by the number of independent parameters of the associated Hamiltonian.

The information manifold can be thought of as a "mesh" on which the common physical and thermodynamic notions of matter, energy, temperature, and so forth, exist, interact, and may be observed. It is natural to consider both the geometric and topological properties of this manifold and ask what connection these have to the world described by the given probability measure.

With the Natural Theory of Information, we now have a way to measure complexity. This will be instrumental in the development of the Theory of Complexity and Emergence, which we discuss in the next chapter.

Chapter 3 – Complexity and Emergence

Figure 3.1 Examples of polyhedra with their Euler characteristic.

3.1 Complexity

The concepts of complexity, emergence, and complex adaptive systems have eluded mathematical definition since the 1980s (Kolmogorov, Gell Mann, Santa Fe Institute, etc.). These terms primarily had appeared in the social sciences and biology where they are used colloquially often to describe flocks of birds, small world networks, phenomena where the whole is more than the sum of its parts, or systems that seem best understood in terms of their own rules rather than rules of their parts. But many efforts have been made to make these terms precise. In mathematics, information theory, and computer science, scientists make use of Kolmogorov's notion of algorithmic complexity, Murray Gell-Mann's effective complexity, and others. But the gulf between worlds remains, as these precise notions have famously failed to capture a satisfactory amount of the intuition that underlies the colloquial usage.

We have created definitions of both complexity and emergence, shown how these definitions capture (and in fact refine) the colloquial notions, and then developed a mathematical theory that clearly captures those definitions. This theory has a remarkably broad scope of applications in emergence and evolution of life.

We define complexity as a real valued invariant of an information-theoretic system that intuitively captures the above ideas. This invariant can be motivationally derived from the principle that the whole is more than the sum of its parts. If a complex system is one in which the whole is more than the sum of the parts, then we may take the measure of its complexity to be how much more information the whole possesses than the sum of its parts. Γ, our measure of information (either classical or quantum), was defined to be how much more information the whole possesses than the sum of its parts, and we then define complexity of the system to be the expected natural information content.

While this effectively captures ideas such as the whole being more than the sum of its parts, it fails to capture the inherently binary concepts underlying complexity and emergence, such as whether or not a system acts in accordance to its own rules or should be considered an individual entity.

This will be addressed by our definition of emergent systems, and a refinement of this notion into an integer-valued invariant we call *emergent complexity*.

We will take as our motivating concept the idea of an emergent system as a system that is meaningfully *its own entity*, so that it may, for example, follow a set of rules that may seem entirely different from those that govern its components. Intuitively an entity may be thought of as a collection of all the "correct" components. A molecule is a collection of all its atoms (and no others). A flock is a collection of all the birds in the flock. But what defines the flock as an emergent entity is the interdependence of the individual birds. The molecule is an entity, but its constituent atoms in an unbound state are not. The flock is an entity, but the collection of its constituent birds, when they are isolated, is not. So, to characterize an entity, we must capture a system whose parts are interdependent, and which contains exactly those parts that happen to be interdependent, and no others.

Complexity, or expected natural information, is a measure of the average statistical interdependence of the components of that system. Therefore, the relevant components of a system are precisely those that increase its complexity. This perspective motivates our definition of emergence to be at a local maximum of complexity.

3.2 Emergence

We define *emergent system*s to be those systems which are maximally complex, and an integer-based invariant of this system becomes its *emergent complexity*. With these definitions of emergence and our definition of emergent complexity, we now capture the above colloquial intuitions. This definition of emergence centers on a single trait of the colloquial concept—in this case, a hierarchy of construction, but it can be shown that the other desired traits follow from this definition.

The emergent complexity then can also be given as the Euler characteristic of the terminal homology theory applied to the information manifold, thereby connecting complexity, homology theory, and the information manifold.

Name	Image	Vertices V	Edges E	Faces F	Euler characteristic: $V - E + F$
Tetrahemihexahedron		6	12	7	1
Octahemioctahedron	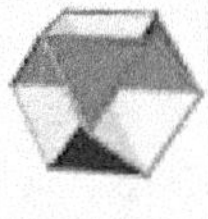	12	24	12	0
Cubohemioctahedron		12	24	10	−2
Great icosahedron		12	30	20	2

Figure 3.1 Examples of polyhedra with their Euler characteristic.[6]

It can be shown that emergent structures maximize predictivity. As an intuitive example, if you are given a set of a trillion atoms and asked to predict their future behavior with maximal efficiency and minimal error, you would be relieved to find out that those atoms belonged to, say, a single ball on a pool table, or to the body of a hungry man headed to his favorite restaurant, as these systems are vastly more "descriptively efficient" than their components taken independently.

3.3 Complex Adaptive Systems

Complex Adaptive Systems (CAS) are defined as those systems that are emergent and persist in their structure over time. Molecules, organisms, or an army of ants are all examples of complex adaptive system and will be discussed in further detail later in this book.

Given this, we have proven that emergent complexity is an invariant of complex adaptive systems. This essentially shows that the notion of complex adaptive systems is best described in its own terms, i.e., the whole follows a different set of rules then the parts.

One commonly claimed property of complex adaptive systems that is conspicuously absent in our definition is that of phase transitions. The fact that our complex adaptive systems lack phase transitions is largely the same as the fact that emergent complexity is constant for a CAS. There is a slightly more general notion, which we call a weak CAS, which is a CAS but without the condition that the best predictor map γ be continuous. It is then a theorem that a weak CAS has a best predictor map with at most countably many discontinuities, each of which is isolated. These points are intuitively the phase transition locations, which a (non-weak) CAS doesn't possess, being more robust over time. In weak CAS, it can be shown that there is a $t0$ that can be thought about as corresponding to the point of death and therefore the point of transition between when the system is best understood as a human, and when it is best understood as (say) only a collection of cells.

Chapter 4 – Laws of Thermoinfocomplexity

4.1 First Theorems

4.2 Punctuated Equilibrium

4.3 Other Applications

Figure 4.1 Gradualism vs. Punctuated equilibrium.

4.1 First Theorems

The theory of Thermoinfocomplexity states that life and other complex adaptive systems arise due to the same pressures of statistical mechanics by which the Second Law of Thermodynamics, and other theorems in that setting, derive. Several authors, perhaps most famously Arto Annila and Jeremy England, have made this observation with various degrees of mathematical and empirical support, holding that complex adaptive systems are maximally efficient energy dissipators, or are entropy maximizers. We have been able to prove precise versions of these claims, and in such a fashion to provide empirical application.

Theorem 2:

(First Fundamental Theorem of Thermoinfocomplexity)
In a thermodynamically-closed system (O, S, P(t)) with

a

continuous energy gradient whose magnitude is bounded
below by a suitable constant at time t0, Γ 0 s (O(t0)) > 0

The "suitable constant" is determined by the size of the system, its entropy, and the magnitude of the energy gradient.

Thereom 2 basically states that complexity must increase in systems where there is a suitable energy source. We claim that the Earth together with the energy gradient induced by solar radiation satisfies these conditions and can prove this contingent on reasonable assumptions.

In a complexifying system, "most" CASs (as a weighted average) in the system are themselves complexifying. Since their complexity is increasing, they must be increasing towards a local maximum.

Theorem 3:

In a complexifying system, emergence attracts; it is in a basin of attraction, i.e., at a local energy maximum (Second Fundamental Theorem of Thermoinfo-complexity).

The above theorem describes the property of complex adaptive systems that they persist, in some sense, over time. Here adaptivity is rather ambiguous, but could be taken to mean simply that the system generally changes over time. Relatedly, it is sometimes said of complex adaptive systems that they exhibit robustness, which may be a more precise interpretation of adaptivity, and a property which is reflected in the fact that—as basins of attraction are open sets, i.e., they reside at a local energy maximum—if the CAS is perturbed by a small amount, it will remain in the basin for its best predictor, and all else equal, tend again towards equilibrium.

4.2 Punctuated Equilibrium

Punctuated equilibrium is the theory that the evolution of species proceeds in characteristic patterns of relative stability for long periods of time interspersed with much shorter periods during which many species become extinct and new species emerge. Although this phenomenon is observed in our evolutionary history, we have not been heretofore able to prove it. Now, with the mathematical tools we have developed, punctuated equilibrium can be proven.

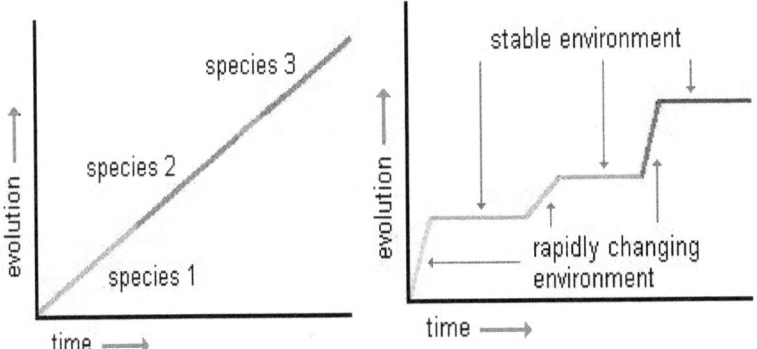

Figure 4.1 Gradualism vs. Punctuated equilibrium.[7]

We demonstrated in the above section that in a complexifying system, the average complex adaptive system will trend towards a local maximum of systemic complexity, and thus a state of emergence.

The proof of punctuated equilibrium also leads us to the Third Theorem of Thermoinfocomplexity.

Theorem 4:

(Third Fundamental Theorem of Thermoinfo-complexity). In a complexifying system, $\Delta\Gamma e(O) \geq 0$.

Theorem 4 is stating that, for a system that is complexifying, emergence will increase.

4.3 Other Applications

Metabolic rate, heart rate, lifespan, and many other physiological properties vary with body mass in systematic and interrelated ways. This idea is called Allometric Scaling. It is well-known that there are many log-log relationships between the mass of an animal and various other properties, most famously the basal metabolic rate of the animal. Over the past few decades, Geoffrey West has improved the scope and understanding of these relationships, showing that the associated power law can generally be derived under certain modest assumptions from the fact that our universe has three spatial dimensions, and that the correlation between mass and

58

metabolic rate is improved if one looks at the correlation instead between the fractal dimension of the animal's circulatory system instead of mass. Here the log-log linear relationship has improved variance, but there are still outliers, perhaps most famously humans. We argue that we can improve further on this relationship by correlating systemic complexity density to energy efficiency. Intuitively, the more tightly-networked the system is—and thus the greater its systemic complexity—then the more it can distribute energy without waste. From this description, it may be plausible that the fractal dimension of the circulatory system correlates to its systemic complexity density, but a human's disproportionately-intricate brain could explain why the fractal dimension of our circulatory system predicts a significantly lower basal metabolic rate than given by observation..

From the proof of the above theorem, we can obtain the following:

Theorem 5:

An ECAS operates at a local maximum of energy efficiency.

Here we emphasize that this does not refer to a maximum with respect to time, but rather with respect to the space. This establishes one of the several core claims that motivated the theory of Thermoinfocomplexity.

Lastly, using the machinery of Thermoinfocomplexity developed thus far, we can show the following.

<u>Theorem 6</u>:

Every ECAS has a finite operational temperature range, and in a complexifying temporal system the weighted-average ECAS will run "close to the top" of this range.

Mammals, being made essentially of water and all of the same proteins, have a potential operating temperature window of between 0° C, below which water freezes and mobility is impossible, and 41° C, above which their constituent proteins denature. Within this range a priori all temperatures are permissible; yet mammals run hot, with standard body temperatures between 35° C and 40° C with only a very few exceptions. Theorem 6 proves why.

We have now demonstrated the fundamental principles of Thermoinfocomplexity:

- Systems in the Universe are complexifyng.

- Emergent complex adaptive systems are robust and persist over time.

- Complex adaptive systems will emerge to higher states of complexity.

- Emergent complex adaptive systems operate at a local maximum of energy efficiency, hence such systems will be selected to persist by the principle of least action.

We show that the interplay between emergence and energy allows for a substantial statistical improvement on Darwin's theory of evolution, which describes the statistically-likely trend towards greater complexity in any system in a suitable energy gradient, living and non-living alike.

Upcoming chapters will explain Gibbs free energy flow and discuss the application of the theory to the emergence of life, multi-cellularity, social networks, civilizations and superorganisms.

Chapter 5 – Gibbs Free Energy Flow

5.1 Gibbs Free Energy

5.2 Gibbs Free Energy Flow

Figure 5.1 Edward Lorenz's three-dimensional mathematical model of a strange attractor.

Figure 5.2 Glucose and glycogen entrained metabolic cycles.

5.1 Gibbs Free Energy

The chemist J. Willard Gibbs wished to describe why certain chemical reactions occur spontaneously, while others require an input of external energy (usually in the form of heat) in order to occur. He described reactions which need an external heat source to occur as "endothermic," and reactions which occur spontaneously and give off heat as "exothermic."

To quantify these forces, he introduced the number we now call Gibbs free energy. In a system representing a substance at a certain temperature T, pressure p, and volume V, the equation for Gibbs free energy as we would write it is:

$$G(p,T) = U + pV - TS$$

where U is the "internal energy" of the system, and S is the entropy. The Gibbs free energy represents the total available energy in the system. It subtracts the energy that cannot be used because it is bound up in chaos. He discovered that reactions in which the end product had a lower Gibbs free energy than did the same system before the reaction, were exothermic; they occurred without the need for any additional input of energy. We can therefore generalize the concept of Gibbs free energy beyond its original context and use it as a description for the amount of energy in a system which is available to do work on another system.

By changing our focus from energy to Gibbs free energy (GFE), we can develop a better understanding of how systems are able to make use of the energy contained in their environments. For a system to perpetuate itself over time, it must perform a variety of crucial functions. These functions can include any number of different things, from the repair of DNA in cells, to movement in the human body. For these reactions to occur with any degree of reliability, they must be exothermic, and so the thermodynamic system must have a high GFE, such that the GFE of the end-state of the system after having performed these functions is lower than GFE at the beginning-state.

Consider a sugar-consuming organism mentioned previously. For this organism to engage in its various vital processes, it needs to rearrange itself in order that the reactions are exothermic, or at least not endothermic. To do so, it needs to raise the amount of Gibbs free energy it contains within itself and its structure. If we view the organism and the sugar as a single thermodynamic system rather than as two separate systems, the digestive process is one in which it transforms both the sugar and itself to produce ATP, concentrating its own energy in a low-entropy form while leaving a high-entropy waste product. This process, if we look at the yoked thermodynamic systems as one system with one quantity for total GFE, will lower the GFE of the combined system.

In other words, once the organism has fed, the waste products are viewed as two separate systems again, and the GFE is distributed differently between the two. The GFE of the organism is now higher than the GFE of the hungry organism, and the GFE of the waste products is much lower than the GFE of the uneaten sugar. Thus, though some total GFE was lost in the interaction between the organism and the sugar, the GFE was redistributed such that the proportion of total GFE in the organism subsystem is higher than it was before the interaction. This is the manner in which we might say that one thermodynamic system can "consume" the Gibbs free energy of another system, and throughout this book we will speak in these terms to simplify our discussion.

Of course, not all of the basic functions of life are exothermic. Photosynthesis, for example, is a process by which plant cells use the excess energy in their environments, in the form of sunlight, to create the sugar glucose. Here, they take the free disordered energy in their environment and bind it in a low-entropy form: glucose, which can later be disassembled as a source of energy for other reactions. This can be thought of as transforming a situation of low Gibbs free energy into one with higher Gibbs free energy, through an exothermic reaction (due possibly to the external source of energy); it's also another way in which a system consumes GFE from its environment. The great utility of the concept of GFE is that it subsumes both the free disordered energy and the low-entropy organized energy under a single rubric.

The Second Law of Thermodynamics tells us that, in a closed system, entropy increases, and Gibbs free energy therefore dissipates. Energy enters an open system from outside (in a plant, for example, in the form of sunlight) and through a series of reactions, eventually dissipates as entropy. In Thermoinfo-complexity, we will describe any process of energy transfer as the flow of Gibbs free energy. The law by which reactions occur spontaneously, if they lower GFE, will be treated as the fundamental axiom governing the dynamics of whatever system we study, and we will examine all of the complex behaviors that can arise from this simple rule.

Here it is important to specify that the law by which reactions occur spontaneously, and which lower Gibbs free energy, is not a deterministic law but rather a stochastic one, describing stochastic systems. By a stochastic system, we mean a system which changes its state over time through a process of cumulative probability, and which means in turn that the future is determined by a combination of its present state and an element of probability. The laws which govern a stochastic system will themselves not be true in a deterministic sense; they will, rather, describe which events are more likely than others to occur. When we say that reactions occur with a lower Gibbs free energy, what we mean is that although at any given point in time the system may change in ways that raise and lower its Gibbs free energy, developments in the system which bring it to an end state in which the Gibbs free energy is lower than it was in the beginning state, are more likely to occur than are developments in which the GFE is higher at the end than it was at the beginning. As this takes a long time to state, we will generally give laws in deterministic form, with the understanding that since we will be dealing with stochastic systems, they should be interpreted in that context, albeit with an inherent dose of probability.

5.2 Gibbs Free Energy Flow from Simple to Complex Adaptive Systems

From simple rules like the tendency of Gibbs free energy to be expended, incredibly complex and beautiful structures can arise. How this happens is the central object of study in Thermoinfocomplexity, and in order to examine it we must combine our understanding of thermodynamics and information theory with complexity theory. Complexity theory studies emergent properties of a system, behaviors, and characteristics which are not apparent when looking at the system on the level of interactions between individual constituent parts but emerge when we view the system as a whole.

Consider, for example, the snowflake. Snowflakes are nothing more than frozen water, accumulations of solid H_2O molecules. They are produced by the random aggregation of ice crystals in the atmosphere. However, most snowflakes (here we are dealing in probabilistic, statistical facts) have a hexagonal symmetry: the same pattern repeated six times. This derives from the structure and shape of the H_2O molecule, though there is no hint of any such symmetry in the molecule itself. The symmetry is an "emergent" property of large clusters of water molecules at low temperatures. Symmetric structures are very low-entropy. If we think in terms of the information necessary to describe an object, a hexagonally symmetric object can be described in full by describing only one-sixth of the whole; the rest can be extrapolated. How can a low-entropy structure arise from a random collision of water molecules, a high-entropy situation? It may seem at first to violate the Second Law of Thermodynamics, but we will show throughout this book that the emergence of low-entropy structures in high-entropy situations is in fact common as well as wholly reconcilable with classical thermodynamics.

To understand this phenomenon, we should first discuss the idea of a "catalyst." We discussed earlier how reactions within thermodynamic systems which lower the Gibbs free energy of the system, will occur spontaneously. However, some reactions will have middle states between their beginnings and their ends, and these have a higher Gibbs free energy than both the beginning and end states. Thus, some additional input of energy into the system is required in order for the reaction to occur, even though it is exothermic. This additional energy is called "activation energy." A "catalyst" is anything which is introduced into the system that lowers the activation energy required for the reaction, thus allowing it to occur with less external energy. In the classic chemical sense, it allows the reaction to occur at lower temperatures. In a statistical sense, when reactions are actually composed of thousands or millions of smaller reactions on a molecular scale taking place in a system containing a very large number of molecules, the catalyst has the effect of causing large-scale reactions to occur more quickly.

One of the most important biological examples of catalysts are enzymes. Some enzymes, called ribozymes, are made of RNA. Ribozymes have been discovered that are able to self-catalyze: these strands of RNA act as catalysts for the synthesis of copies or portions of themselves. More generally, in biological systems, there exist "autocatalytic sets" of enzymes. RNA-based enzymes catalyze the creation of certain protein enzymes, which then catalyze the creation of the original RNA-based enzymes. There may even be more enzymes in this process, so long as it is circular, eventually leading to an enzyme which catalyzes the production of the first enzyme in the set. (See Chapter 6 for a more detailed examination of this example.)

From the scale-free viewpoint of Thermoinfo-complexity, we can view an autocatalytic set of enzymes as a single object, and, should we choose to do so, discuss large populations of autocatalytic sets. We can also take this concept even further outside of the biological context and discuss "autocatalytic structures," by which we mean any distribution of energy, or pattern of energy flows, which causes it to be more likely to exist in the future than it would be to arise spontaneously.

An excellent metaphor for this phenomenon is rain falling on a mountainside. The path which the water takes is determined at each point by the direction of steepest and easiest flow. Although we cannot predict exactly the path each individual molecule of water will take, we can describe the various streams and channels by the volume of water that will pass through them, and we can say at what speed it will flow. This process is stochastic; we can give probabilities for what each molecule will do, and there are so many molecules that they can be treated as a continuous mass rather than as a very large group of individual objects. The pattern of water flow is analogous to a pattern of energy flow in our generalized perspective of Thermoinfo-complexity. In this light, we may think of photons and other forms of energy going through the Earth's atmosphere, encountering various forms of matter, and slowing down, but finally finding their way into the vast expanding universe as per the Second Law of Thermodynamics.

As water flows down the mountain, it erodes the surface over which it flows. This continuous stream has the effect of deepening and smoothing the channels it follows, making it even more likely that water will flow through these paths in the future. Thus, if the water follows a certain pattern of flow, it is more likely to follow the same pattern (or a very similar one) at the next moment in time. Thus, this pattern is an "autocatalytic structure." It makes itself more likely in the future, and we can then look at it as a persistent object. It is a central point of Thermoinfo-complexity that most emergent properties of large systems can be viewed as autocatalytic structures.

Like our notions of information, our discussion of complexity above is deeply tied to randomness and chaos. The irregularities in a system make it more complex, and these often arise because a system is governed by rules, which are not exact but probabilistic. All of the systems we discuss in this book, and all of the systems approachable through Thermoinfocomplexity, either because of limits to effective measurement or because of the inherent nature of the systems will to some extent be probabilistic as well as chaotic.

"Chaos" refers to the uncertainty of predicting what a dynamical system will look like in the future because of powerful effects from small differences in initial conditions. Chaotic systems need not be random or stochastic; they may be entirely deterministic, albeit very difficult to predict. However, if we lack the ability to determine the conditions which govern it exactly, any deterministic system can be viewed as a stochastic system.

Instead of describing the exact configuration of a system at a given point in time, when describing chaotic systems, we speak in terms of "attractors." An attractor is a set of states to which the system will tend to be attracted. There are several types of attractors, the description of which will make this definition clearer.

The simplest is a "point attractor." A point attractor is just a single state of the system. If the system describes an object moving in space, this is a specific point in space. The requirement that an attractor be fixed means that if the system is at this point, then as time moves forward it will not change. The nearness means that if the system is near to this point, or in a state similar to the attractor state, then as time moves forward it will change so as to become closer or more similar. In thermodynamic systems, a point attractor is often a state of lowest Gibbs free energy, or at least a point that has lower Gibbs free energy from all nearby points. Due to the general law by which thermodynamic systems develop, the system will evolve in time to become closer and closer to this point.

Another type of attractor is a "limit cycle." This is a one-dimensional closed loop in the space of states that the system can take and follow repeatedly over a fixed period. If one considers an ideal pendulum, the positions and momentum it takes as it swings back and forth form a limit cycle. Limit cycles are regular patterns that a system can take. Once a system is in any state contained in the limit cycle, it can never leave that limit cycle.

Attractors can exist even if points nearby the attractors never actually reach the attractor. The system can spiral closer and closer toward a point attractor without ever reaching it, and a system can also wobble forever in a pattern that is very close to a limit cycle, but never actually fall within the limit cycle.

Attractors can have a great number of different shapes, including ones exhibiting the strange properties of non-integer dimensions. Attractors with fractional dimensions, "fractals," are referred to as "strange attractors." These are difficult to construct using top-down mathematical methods, but they frequently occur as emergent phenomena in systems that are described as the results of many interacting systems. A famous example is the Lorenz Attractor, which describes pathways taken by an object moving in space following a few ordinary differential equations meant to simulate atmospheric convection.

Though existing inside three-dimensional space, the attractor itself has a dimension estimated to be near 2.05.

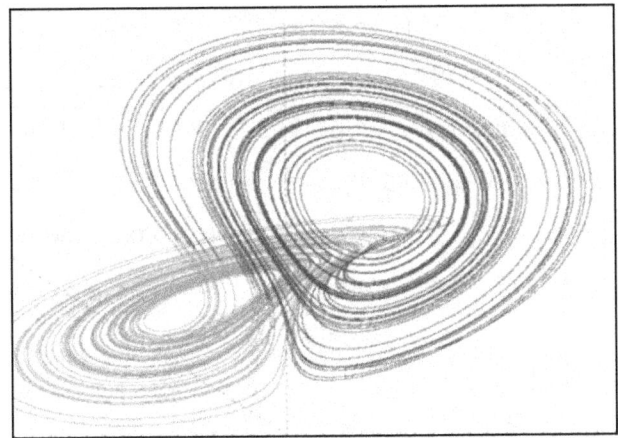

Figure 5.1 Edward Lorenz's three-dimensional mathematical model of a strange attractor.[8]

These dynamical systems can develop from self-interactions in various ways. One of the most important is through feedback loops. Positive feedback loops occur when an event makes itself even more likely to occur in the future, or likely to occur more intensely. Uncontrolled population growth (given unlimited resources) is a good example of a positive feedback loop. As the population grows, its rate of growth increases at an endlessly accelerating rate.

This endless positive feedback loop is, of course, not a realistic situation, and as such it does not occur frequently. Combinations of positive and negative feedback loops occur more often in nature. A negative feedback loop occurs when an event makes itself less likely to occur in the future, or to occur less intensely. A good example from biology is the cycle of the hormone insulin and glucagon in the human bloodstream.

When we eat something with sugar, that sugar (disaccharide) is broken down into two glucose molecules. In the small intestine, the glucose is absorbed into the bloodstream and is then diffused into the spaces between the cells of the body. There, it is attracted to certain receptor proteins on the cell's surface. It then passes through the cell membrane and enters the cytoplasm. In all cells, this starts the process of ATP production. The ATP within the cell increases to a certain threshold, where the potassium channels on the cell membrane close and the calcium channels open, resulting in an influx of calcium across the cell membrane. This ion transfer changes the electrical balance between the inside and outside of the cell membrane, in a process known as depolarization. In this process, the change of the environment in the interior of the cell becomes an attractor for the calcium ions. In the pancreas, calcium follows this attractor pathway. When it reaches a certain level, the pancreatic cells begin to release the hormone insulin into the bloodstream.[9]

In the blood stream, insulin is attracted to proteins on the surface of all cells. Once insulin attaches to those proteins, a chemical reaction occurs in the cell membrane that allows the cell to take in glucose from the bloodstream. When glucose enters the cell, it is used to generate ATP, much as it does in the pancreatic cells. However, this process cannot go on forever, as the glucose level would build up to a toxic level. As the level of glucose in the bloodstream increases, insulin levels drop through a negative feedback loop that affects its synthesis in the pancreas, and the excess glucose passing through the liver is converted to glycogen. Glycogen is a highly compact but unusable form of sugar which acts as long-term storage. Though not immediately useable, it may be broken down at a later point. Eventually, the surface proteins on the cells that hold onto the insulin let go, releasing insulin into the blood stream, where it is later broken down in the kidneys, or ingested by the cell itself. This interaction between the receptor and the insulin is transient, and the cell's intake of glucose is transient as well. When no additional glucose enters your cells, the ATP levels slowly start to fall. In the pancreas, this causes the calcium channels in the insulin-secreting cells to close and the potassium channels to open, so that insulin is no longer released in large bursts.[10] This is a type of positive feedback loop. When glucose is high, insulin levels are high as well. But when the signal of glucose disappears, the insulin disappears with it. These two processes are entrained like the interlocking of the gears in a clock.

Insulin also has a negative-feedback effect on the hormone glucagon, which stimulates the breakdown of glycogen into glucose molecules, releasing glucose into the blood when blood sugar is low. A drop in glucose level (due to its use by the cells and conversion into glycogen in the liver) shuts down the release of insulin. With insulin no longer being released, its concentration in the blood drops, while glucagon levels increase as per a negative feedback loop. The reduction of one allows for an increase in the other. Glucagon turns stored glycogen in the liver into glucose and releases it into the blood. Blood sugar increases, and the cycle whereby insulin is released, and glucagon lowered, begins again. Insulin and glucagon cycle in this manner, with the one being predominant in the blood to decrease blood sugar, followed by the other in order to raise blood sugar. Together, these two hormones work to keep blood sugar at a relatively constant level in order to maintain homeostasis in the body. Because increases in insulin levels decrease glucagon levels, this hormonal release constitutes a negative feedback loop. When insulin levels fall, glucagon levels increase once again, allowing the cells to put their stored sugar to use. As insulin levels fluctuate due to this pancreatic rhythm, so too do the levels of glucagon. In this way, the insulin-glucagon cycles are linked with one another and therefore entrained together. Cells can use this glucose to manufacture ATP. ATP produced by this process is free to power other cycles, like the contraction of the heart muscle, which pumps the blood that carries glucose from the small intestine to the pancreas.

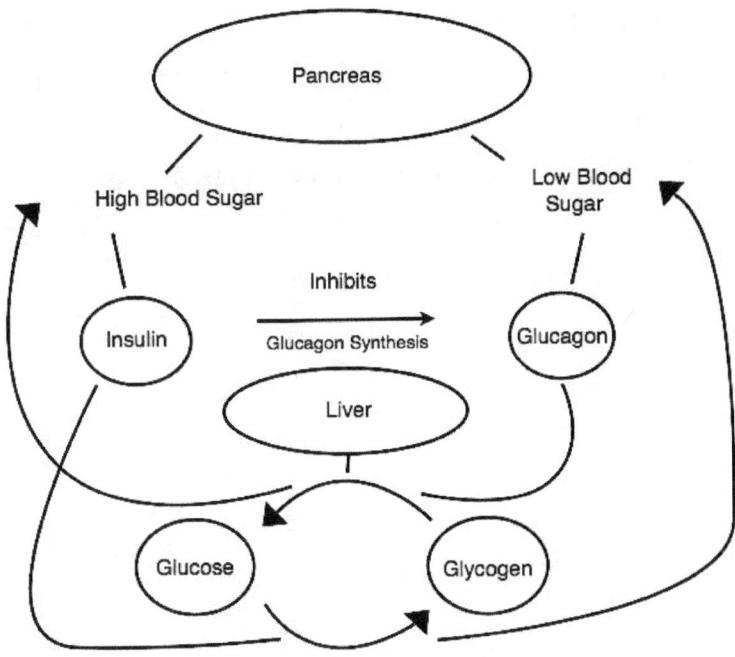

Figure 5.2 Glucose and glycogen entrained metabolic cycles.

This cycle is an emergent phenomenon arising from the basic rules of how insulin and glucagon affect cells.

Thermoinfocomplexity is a scale-free general theory that describes the emergence and evolution of various forms in nature that result from stochastic encounters of energy and matter in the cosmos. Specifically, in this book we describe its application to the emergence of complex adaptive systems in terms of the flow of Gibbs free energy and its conversion to cybernetic information within the structure of complex adaptive systems, all the while obeying the Second Law of

Thermodynamics. The theory is scale free because it allows the emergent self-similar structures to be viewed from a higher or lower perspective using the concepts of microstates and macrostates. We intend to show the specific application of this theory in a number of different situations throughout the remainder of this book.

Chapter 6 – Life's Origins

water (H_2O) and releasing ammonia (NH_3)

It was an early spring morning. I was walking on a narrow trail in the mountains of Kurdistan. I sat by a small stream pouring into a very small pond. Sunbeams were visible in the vapors rising from the melting snow. It felt like the photons were talking to the snow's ice crystals, telling them to wake up. Ice heard the call. Yawning, it became water then rose as vapor mist. The vapor mist made the scaffolding for the sun's rays to talk to my eyes. In awe of this vision, I threw a breadcrumb in my hand into the small pond. Suddenly, a tiny fish appeared from nowhere, nipped at the crumb, and before long at least ten little fish were nipping at the bread. How did the bread send the fish the information of its presence? And how did the fish communicate it to each other not to collide in their nipping at the bread? Oh, what a spectacle, I thought! Nature is buzzing with myriads of energy and information exchanges. In that quiet morning, I heard the bass-bassoon of nature's humming conversation from water molecules to the clamoring fish. To my senses in the chilly mountain air, it was as loud as the hum of the energetic traders of information on the floor of the New York Stock Exchange. Then I fell back to daydream about my life on this planet and wondered where it all began.

6.1 Life's Building Blocks: Oparin and Haldane's Hypothesis

Where did life come from? In the beginning there was a big bang. Then there was the sun and the Earth. Then inorganic material became organic material. Then organic material organized itself into the first life forms. Life has been around for 3.5 billion years of Earth's 4.5 billion-year history.

But, how did life emerge? Charles Darwin laid the groundwork for a unified theory of biology when he showed that all life unfurled from an ancient, simple origin. His theory of natural selection helped explain changes in life forms over time. But how did life begin? How did the origin of species actually take place? Two 20th century scientists, Alexander Oparin and J.B.S. Haldane, hypothesized that organic molecules—the precursors to life—would be favored to emerge from simple inorganic building blocks. They believed that the entire process could have occurred in Earth's early environment. If properly tested, their ideas would illustrate a plausible route toward the origin of life.

Oparin and Haldane recognized that inorganic building blocks would only give rise to organic molecules under the proper conditions. Not every environment of ancient Earth would have been suitable for the emergence of life. Charles Darwin had wondered what would have been the appropriate conditions for life's beginning in a letter to his friend, botanist Joseph Hooker. He was probably conjecturing what we think of as the "primordial soup."

If (and Oh! What a big if!) we could conceive in some warm little pond, with all sorts of ammonia and phosphoric salts, light, heat, electricity, etc., present, that a protein compound was chemically formed ready to undergo still more complex changes…

Such a warm little pond would contain all the simple, inorganic chemical precursors to life. It would require only an energy source—lightning, perhaps—to provide the energy needed to initiate chemical reactions. The ingredients were there, but they had to be made to communicate. We should then see the emergence of organic compounds, which form the basis of life as we know it. Organic compounds—life's building blocks—are really not so different from nonliving material, after all.

6.2 Chemical Evolution

Both living and nonliving material are made up of the same constituent parts: atoms. Atoms are made up of smaller constituent parts—protons, neutrons, and electrons. Protons are positively charged and neutrons are neutral. Both of these particles are found in the nucleus of an atom, while electrons are negatively charged and spin around the nucleus within specific areas, known as atomic orbitals. Protons and electrons are attracted to each other electrostatically, much like the opposite poles of a magnet. In this way, protons repel protons, and electrons repel electrons.

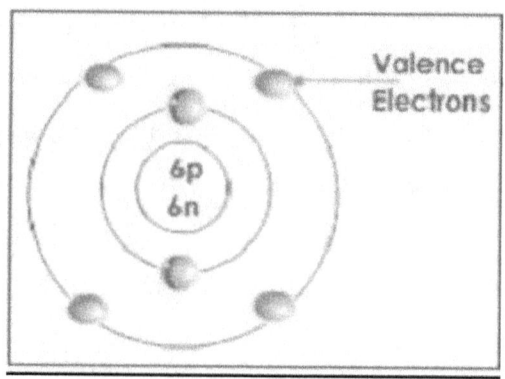

Figure 6.1 Carbon atom, with its nucleus containing six protons and six neutrons, two electrons on the inner shell, and four valence electrons on the outermost orbital

In an atom, the electrons closest to the nucleus are tightly held and quite stable. But electrons on the outermost orbitals—the valence shells—are partially shielded from their nucleus by the electrons between them, making it easier for them to be pulled by the protons of other adjacent atoms. When two atoms come close to one another, the protons of one will attract the electrons on the valance shell of the other. Keeping the two atoms apart would require energy, just like holding magnets of opposite poles apart. But if the atoms get too close, the protons from one atom will start repelling the protons of the other atom, and it will take additional energy to push them any closer, like forcing two magnets with the same pole together. The atomic bond length is the distance at which the valence shell electrons (*Figure 6.1*) can be close enough to both nuclei to share their electrons without the protons in either atom repelling one another.

This length represents the state at which the energy is lowest, that is, when it takes the least energy for the electrons to occupy their valence orbital. It is this more stable state that limits the area through which an electron can travel, as both nuclei are exerting an electrostatic pull to keep the electron within a certain distance. In a non-bonded atom, an electron could be found anywhere within a specific orbital of that atom. But, when an electron is contained in a bond between two atoms, the equal pull causes it to be located more precisely between the two atoms, giving us more information about the electron's location. In other words, the two protons are point attractors for the electron, creating attractor paths for the shared electrons. Because the electrons, as well as the two protons, are subject to being pushed and pulled by the other atoms around them, this path is not a line, but rather it is many lines within a range, known as an electron cloud.

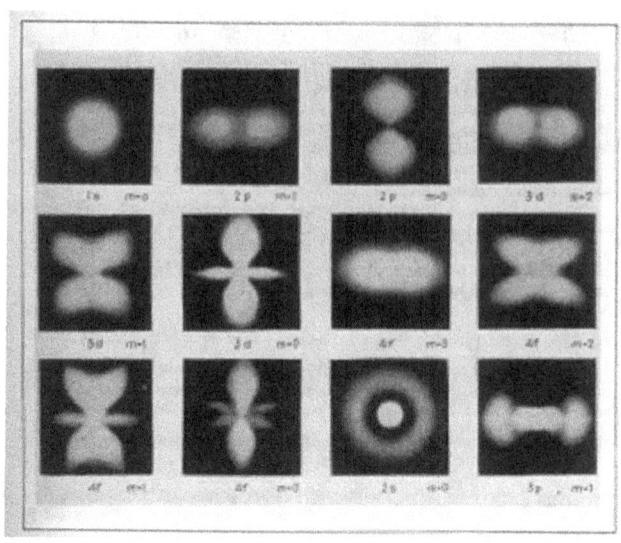

Figure 6.2 Early photographic representations of electron clouds, made in 1934 by spinning wooden models.[11]

Albert Einstein and Neils Bohr theorized that electrons have both particle and wave-like characteristics. Due to the wave-like properties of electrons, the valence shell may be filled to stabilize an atom further. This is done either by atoms giving away or receiving valence electrons, or by sharing the electrons in a bond. In general, when the orbital closest to the nucleus contains two electrons, it is filled and stable. For example, helium only has two electrons in its first orbital, and it is one of the most stable atoms. Orbitals further from the nucleus become most stable—have the lowest energy state— when they are occupied with eight electrons.

Atoms, much like organisms, survive through time by becoming as stable as possible by binding to other atoms through chance encounters. Through bonding, atoms enter their lowest possible energy configurations and contain the highest information content able to be captured in their bond. Once formed, this process is statistically irreversible, unless a large amount of energy is applied from outside the system, in an endothermic reaction. In *Figure 6.3*, a molecule undergoes a reaction, ending in a lower energy state than when it began. It is more stable after the reaction is completed.

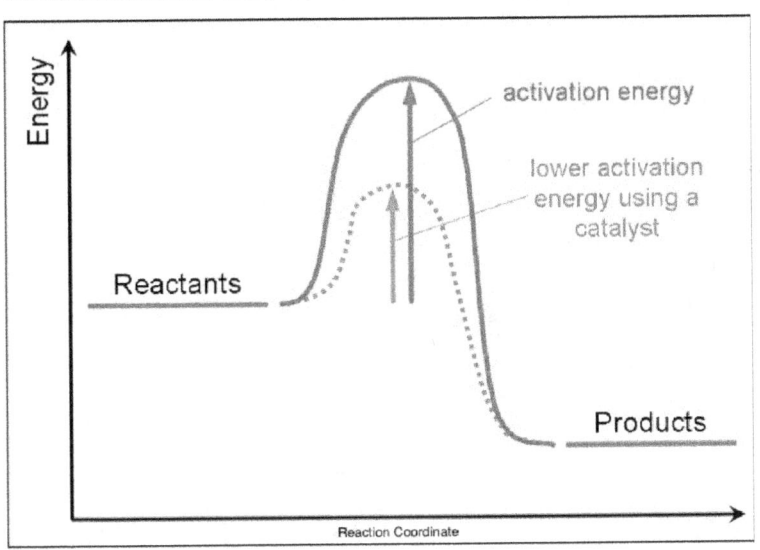

Figure 6.3 Diagram shows an energy releasing reaction. The molecule formed ends in a lower energy state than where it began. The lower curve shows that a catalyst will decrease the activation energy needed for the reaction to occur.[12]

Bacteria require nutrients to make and keep their structure stable; similarly, chemicals require a certain amount of "activation energy" to go through chemical reactions and form new structures. When this energy requirement is met and the electrons form a bond, the electrons in the atoms become confined in the structure, trapping some of the energy, while the remaining activation energy is released as heat into the environment. Although one bond between two chemicals can be stable and energetically favorable, the release of energy into the surroundings increases the probability that the surrounding atoms will now have the energy to collide with each other and form chemical bonds. Atoms use the energy from other reactions, sunlight, and electricity to form chemical bonds in the same way that organisms use nutrients to get to a state where they require less energy to survive. These bonds are like communications between atoms, sharing electrons as their message.

Just as organisms compete for resources, atoms—through probabilistic, chance encounters—compete for partners that give them the most stable bonds. A negatively charged molecule will be attracted to a carbon atom that has its electrons partially pulled away, exposing more of the positive pull from its protons. The negatively charged molecule will form a temporary bond with the carbon atom, causing the carbon to momentarily form five transient bonds—a carbon atom can only handle four bonds—a state that lasts for almost no time and is thus called a "transition state." But soon, the negatively charged molecule will form a more stable bond with the carbon, than it could with one of the other groups of atoms attached to it, breaking the less energetically stable bond, and allowing the carbon to have four stable bonds and a lower energy state than it had before (this process is called an S_n2 reaction). Molecules are constantly competing in such a manner, with dozens of combinations forming in nature, until the most stable form for the environment outcompetes the rest and is selected. Here we see chemical evolution at work.

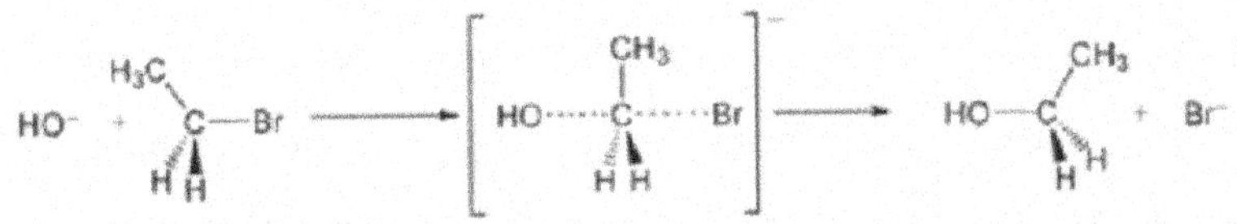

Figure 6.4. Example of an S_n2 reaction. A negatively charged ion interacts with a relatively stable molecule. Note the temporary transition state in the middle. The final, more stable form incorporates the negatively charged ion and ejects a bromine ion.

With more energy input, more chemicals become capable of adding onto a molecule, increasing that molecule's complexity, and thus lowering the energy state of atoms per nanometer and increasing their stability compared to free atoms. The more organized, complex, and stable a molecule becomes, the more information and energy it has stored in its bonds. The more information it has stored, the more energy it releases when the molecule breaks apart. This is the way in which chemicals evolve and are selected for based on stability, gaining more complexity and building on top of one another.

Inorganic and small organic molecules are formed and persist through time with the help of small molecules, known as "catalysts," which bring down the activation energy required to form a new molecule. Catalysts are not used up in the process. For example, iron serves as a catalyst for the synthesis of ammonia (NH_3) from nitrogen (N_2) and hydrogen (H_2), known as the Haber Process. N_2 attaches to the iron particles, which weakens the bonds between the two nitrogen atoms, and then the same happens to hydrogen. With the nitrogen-nitrogen and hydrogen-hydrogen bonds destabilized by the iron, nitrogen and hydrogen electrons within proximity of one another form new bonds, and eventually become an NH_3 molecule. In this way, the strong triple bond in nitrogen is weakened and the hydrogen and nitrogen atoms combine faster than they ever would without iron present. We can think of iron here as a facilitator or a matchmaker, bringing the nitrogen and hydrogen close enough to interact, to communicate through their electrons.

In nature, most of the organic compounds necessary for life, like DNA or proteins, require the help of other large, complex catalytic enzymes, in order to lower their activation energy, allowing them to form quickly. In various experiments, scientists have managed to prove that the precursors to life can form under pre-life conditions on earth, showing a progression from chemical evolution to life.

6.3 Miller and Urey's Experiment

In 1953, Stanley Miller and Harold Urey set out to test the Oparin and Haldane hypothesis. Their experiment would come to be the most famous study into the possible origins of life on earth. In their laboratory, they were determined to recreate the environment of ancient Earth, in which life first originated.

A glasswork apparatus was assembled and sterilized. Water, methane, ammonia, and hydrogen were prepared and sealed inside glass tubes connected by a loop. One chamber contained water, which was heated gently to simulate the hotter earth of 4.5 billion years ago. Another chamber contained a pair of electrodes. These would supply energy to the system via electrical sparks—Miller and Urey's version of lightning. All the necessary conditions had been met for creating simple, organic compounds: the basic components of life. Water provided the medium in which agents (the chemicals) could interact and communicate. The steady flame and sparking electrodes supplied the energy. Inorganic molecules were present in the form of water, methane, ammonia, and hydrogen: H_2O, CH_4, NH_3, and H_2, respectively.

Figure 6.5 Stanley Miller prepares the experimental apparatus.[13]

A week later, the experiment yielded amazing results. Miller and Urey had created organic molecules from inorganic molecules, using only ingredients and energy sources that would have been available in the surface waters of ancient Earth. Between 10 and 15 percent of the system's carbon had reassembled into organic compounds. Some of the carbon— about 2 percent—had even taken the form of amino acids, which are used by the machinery of modern cells to produce proteins. There were basic sugars, lipids, and the precursors to nucleic acids (pieces of DNA), all of which we find in abundance in modern living cells. The system had grown more complex, as if by magic—but it wasn't magic. Organic compounds had emerged from inorganic compounds because they had access to suitable environmental conditions, including an energy source and a container that allowed for the communication of information between simple parts, allowing the formation of an energy efficient, complex network. Though Miller and Urey didn't know it, they were engaging in fundamental complexity research before this theoretical field had even been invented.

6.4 Life from the Bottom-Up

Organic molecules formed from inorganic molecules, in a bottom-up fashion. No one put them together. Amid countless chance molecular collisions, the electrical spark provided the necessary energy to bring some of the inorganic molecules close enough to bind together, enabling the emergence of new, more-complex organic molecules. With enough energy flowing into an open system, a more-complex order had emerged. Miller and Urey's point was that the basic components of life could, in fact, emerge from simple, material origins. The only real division between life and nonlife was in the arrangement of matter and its adaptive complexity that resulted from a hierarchy of complementarities fed by information exchanges between constituent parts, in a network of communication. Energy input, bound in information, and its output, in the form of heat, were the keys to that transition.

The difference between nonlife and life is the evanescence and complexity of the network of relationships, interactions, and information exchanges that are dynamically present in living systems, but less vibrant in nonliving systems.

It is not the case that life arose from some magical process. Life required no special intervention outside of physics and probability. Order comes about from within (with a little energy from the sun). Through stochastic processes, self-organization begets complexity, and eventually, one of those complex physical systems meets our criteria for being "alive," as it begins to interact with the molecules around it and becomes self-replicating. Although here, we should recall, that before life as we know it emerged, self-replication is described by the process of autocatalysis in RNA.[14]

The efforts of Miller and Urey inspired further experiments in later years, when new questions about life's origins arose. They had done important work, but there were still many theoretical alternatives and challenges to be addressed by those who sought to describe the possible origins of life.[15]

Can the origin of life really be reduced to the story of a single primordial soup? There are, after all, many other hypotheses for how life may have emerged. Included among these are the Deep Hot Biosphere, the Clay hypothesis, the PAH (Polycyclic Aromatic Hydrocarbon) World hypothesis, and the Lipid World hypothesis. We won't get into the details of these ideas, suffice it to say that there is no compelling evidence to suggest that life evolved only once or even in the same way, nor did it spread in a top-down manner to virtually all other parts of the globe. In each of the above-mentioned origin scenarios, complex forms arose from simpler ones in the presence of an energy supply, plus an agent (a type of catalyst) that lowers the activation energy of a reaction. None of these theories suggests that a molecule of DNA evolved all at once, with no intermediate steps, and without smaller networks connecting larger ones. It has been a long journey from amino acids to the emergence of lettuce, finches, and chimpanzees. The experimental success of Miller, Urey, and others provides insight into life's origins. At the same time, all origin hypotheses only tell the first paragraph of a much longer story.

6.5 No Compartment, No Life

How do living things, or simply collections of self-replicating organic molecules, persist once they form? All living organisms are composed of either cells or groups of cells. Every cell is bound by a membrane. Most of these membrane bound cells are small enough that they can only be distinguished from their neighbors under a microscope. Even large cells, however, have to keep their internal components in order. No matter how big a cell gets, a membrane keeps the parts of the cell organized, preventing the contents from diffusing in random directions into their surroundings. Without a barrier distinguishing a cell from its environment, a cell wouldn't be a cell. It would lack the necessary structural framework for complex organization. The word "cell" was chosen by Robert Hooke, the "father of microscopy," because it draws on the Latin word for a small room, *cellula*. Hooke was looking through his microscope at a piece of cork, when an image came to him of cloistered monks in individual compartments. Nothing comes in and nothing goes out of each *cellula* without first crossing the barrier of the little room. So, it is with all living cells.

No matter what complicated functions a cell undertakes, its structure would not be possible without its cell membrane. Complex systems form as a result of a network of frequently communicating parts exchanging information. If the nodes in a network are too far from one another, the communication becomes less frequent and more prone to error. A stable structure stochastically arises that favors interactions in reaction cascades and chemical feedback loops that maintain its homeostasis. The cell is no exception, and the cell membrane does more than simply designate a space. It also acts as a gatekeeper, selectively allowing or rejecting the passage of materials and information into and out of the cell. To maintain an internal compartment, a lipid-bilayer, composed of two fatty sheets, separates the inside of the cell from the outside. A cell membrane also contains a variety of proteins embedded in this lipid bilayer. Often these proteins have one end protruding outside the cell and another reaching into the cell's interior.

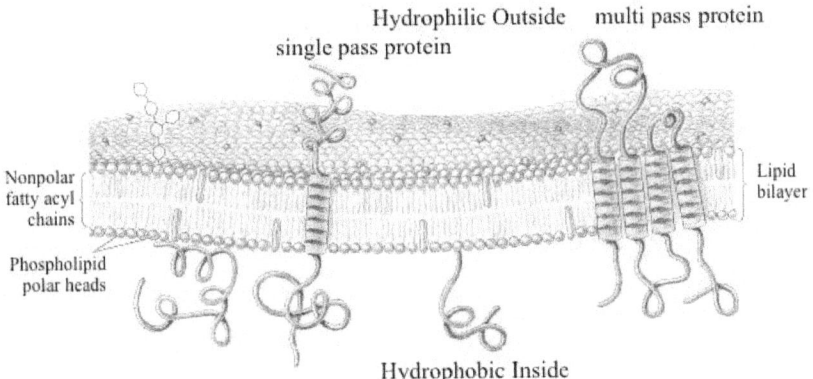

Hydrophilic Outside   multi pass protein
single pass protein
Nonpolar fatty acyl chains
Phospholipid polar heads
Lipid bilayer
Hydrophobic Inside

Figure 6.6 Illustration of proteins spanning the cell membrane (through water channels).[16]

Sometimes, a foreign chemical will be shaped in such a way that enables it to interlock with one of these intermembrane proteins, sticking out from the lipid bilayer. This molecular interlocking triggers a corresponding change in the part of the protein protruding within the interior of the cell. The chemical's shape and the charge of its atoms communicate information about its presence to the cell. When the chemical fits with the protein's binding site, the protein receives a "message," in the form of a cascade of electrical information, manifested as a change in conformation. Such a conformation change is a molecular signal. The chemical that binds to a surface protein changes the electrical charge on the surface of the membrane, and thus communicates its presence to the interior of the cell, which now gains information about which chemicals are in its surrounding environment. Inside the cell, the message is sent through a cascade of chemical reactions, enabling a cell to respond to its environment after it has received the message.

6.6 Chemical Gradients and Retarding the Energy Flow

The most universal feature of the cell membrane is its ability to maintain a chemical gradient. Membrane gradients are capable of doing work, such as distinguishing "good" food from "bad" food, as well as selectively concentrating the "useful" molecules inside the cell. The cell accomplishes this feat through "active transport." By contrast, passive transport is the tendency of concentrated particles to flow along gradients, from high concentration to low concentration. Thermodynamically speaking, active-transport traps energy from the outside, in the form of information within certain molecules, retarding the free flow of energy. In this way, it keeps densely packed chemicals inside the cell, instead of letting them reach equilibrium passively with the outside environment.

The ingenious active-transport pump (a sort of information relay system) is formed from a cluster of proteins embedded in the cell membrane of every animal cell. The sodium-potassium (Na-K) pump is a perfect example of an active-transport pump. The Na-K pump is composed of a protein that uses the energy of adenosine triphosphate (ATP) to bind three sodium ions to the protein part of the pump on the inside of the cell. ATP transfers energy to the pump, while changing into the lower energy-state molecule, adenosine diphosphate

100

(ADP). The Na-K pump changes shape, shuttling the three sodium ions to the outside of the cell. At the same time, the pump binds two potassium ions from the outside of the cell barrier, changing conformation as it brings those ions inside the cell. Once the exchange is made, the pump returns to its starting position and is ready to begin again. The net result concentrates potassium inside the cell and pumps sodium out. The Na-K pump functions by creating a disequilibrium between negatively and positively charged ions across the cell membrane, resulting in an electric membrane potential—allowing the membrane to regulate polarization and depolarization. You may have noticed that for every three charged sodium ions that exit the cell, only two potassium ions enter. This emergent ability to manipulate ionic shuffling across the electrical gradient of the cell membrane, gives the cell a new energetic tool for signaling its electrical state. In other words, the cell uses its emergent polarity, as a means of communicating with its environment. The cell membrane may polarize and depolarize, sending electrical impulses to its cellular neighbors through micro water channels, which can now manifest as nodes in an expanding communication network. Later in the history of life, this electrochemical information transfer system has evolved into what we know as the nervous system. Electrochemical conversation slows the dissipation of energy by binding information in a living-network, giving the cell its "life"—a temporary retardation of energy flow consistent with the Second Law of Thermodynamics.

6.7 Cell Membrane: Lipids, Stability, and Energy Efficiency

The selective, protective membrane barrier of a cell is more- or less-complex, depending on the type of living organism observed. For some bacteria, as well as archaea, fungi, algae, and plants, a second protective layer, known as a cell wall, functions as a second chemical gatekeeper, in addition to the cell membrane. Its chemical composition is more robust than the lipid bilayer membrane, making the cell more durable. The cellulose that makes up the cell walls of plants is what gives these organisms their rigidity (and what makes certain plants difficult for humans to digest). Fungi build cell walls out of chitin, the same tough molecule that builds the exoskeletons that armor arthropods, including crustaceans and insects.

It should be noted that in the history of life, all these cell barriers of varying complexity evolved after the cell membrane. The cell membrane is the historical scaffolding upon which all further cell boundary developments are built. Without the cell membrane, life as we know it would simply not exist. Complexity builds on existing complexity, just as one emergent form relies on the stability of another.

The emergence of a lipoprotein bilayer membrane was extremely valuable in preserving the complex collections of molecules inside the cell during the early history of life on

Earth. Protected from the outside, molecules inside the cell would be more likely to interact and chemically cascade into new molecules, making more of themselves through self-catalysis (a type of reproduction to be discussed in more detail). So now, let's consider some basic properties of lipids, the building blocks of the cell membrane. You are already familiar with some lipids, such as fats, oils, cholesterols, and waxes. From observing these compounds, we can see how they might form the cell membrane necessary for the emergence of early life. When olive oil is poured into a pot of water, the drops of oil that emerge, remain separated. Oil and water don't mix, but oil does mix with more oil. The oil drops tend to form a roughly spherical shape at the top of the water. This happens because oil—a lipid—tends to form the shape that is most energy efficient for its structure: a sphere. Spheres are the most thermodynamically stable forms.

The cell membrane is composed of a double layer of lipids, known as the lipid bilayer. What makes the lipid composition of the cell membrane a "double layer" is that certain lipids called phospholipids (a combination of phosphorous and fat) have two different ends to their molecular structure: one end is hydrophobic (electrostatically repelled by water) and the other end is hydrophilic (electrostatically attracted to it). When these lipids are surrounded by water, they do something remarkable; the lipid molecules arrange themselves into two layers, with the hydrophobic ends of each layer facing inward and the hydrophilic ends facing outward. To reach the lowest energy state, the bilayer spontaneously forms into a sphere.

This spherical lipid bilayer (called a vesicle) is an example of emergent complexity, a precursor of the living cell called a protocell, arising from simple physical laws involving electrostatic attractions.

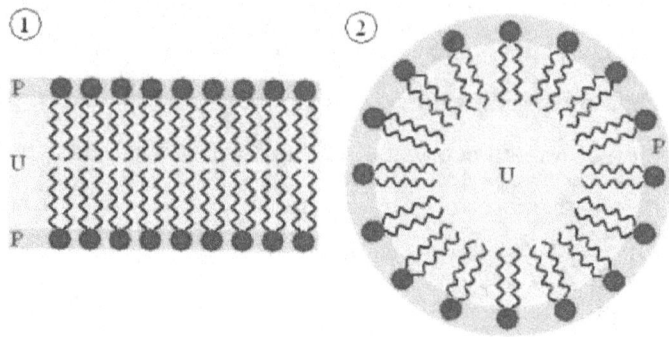

Figure 6.7 Natural formation of lipid bilayers.[17]

The spontaneous formation of lipid bilayers into spherical bubbles (as also seen in soap bubbles) likely provided the initial site for the chemical reactions leading up to the dawn of life. Previously, we discussed the Second Law of Thermodynamics and the battle against entropy that every system wages to maintain homeostasis. Let's keep that battle in mind as we look at the first forms of life. Miller and Urey showed how organic molecules could emerge from inorganic molecules in a laboratory using only a few sparks and the flame of a Bunsen burner. Four billion years prior, the same organic molecules would have needed a relatively protected place if they were ever to be stable and form networks. Since glassware wasn't an option then, a lipid bubble would have done well instead. We can think of such a bubble as a temporary sanctuary for chemical reactions between molecules; the bubble could act like a closed sac, and the organic molecules could be safe and concentrated inside, allowing further reactions to occur. It wouldn't have been a cell, but it was a first step in that direction. And since all this had to happen by stochastic probability and selection of the most energetically stable form, it took place at the scale of millions of years, taking about 3.5 billion years to reach the emergence of bacteria.

6.8 Cell Stability and Catalysts: Building Faster than Falling Apart

One significant barrier to chemical complexity remains: how can we be sure the reaction would have occurred fast enough so that new molecular structures were being generated more rapidly than they were falling apart? For newly formed molecules to persist through time, they would have had to reach a homeostatic state. They would need structural stability. It would be too easy for those chemical products that managed to survive their harsh environment to simply fall apart without their lipid bubble (confinement of cell or cell membrane) shortly after being produced. Gently interacting molecules might simply not have had enough energy to form the most stable product. To form molecules like RNA and proteins, our organic molecules would have needed a catalyst, like the one mentioned in the previous section on chemical evolution.

A catalyst is a substance that increases the rate of a chemical reaction without being incorporated into the product itself. Catalysts work like molecular matchmakers, facilitating new relationships between molecules. By making the transition state—the intermediate molecule between the reactant and product—more stable, catalysts lower the amount of activation energy that must be put in for that reaction to occur. Catalysts don't change the equilibrium or how much overall energy is released or absorbed during the reaction; catalysts only change how much energy it takes for the reaction to go forward. Because the beginning and end products remain the same, any energy gained or released also remains the same. Reactions can still occur without a catalyst, but often at a much slower rate— much too slow to ever be of use within a living organism. Catalysts can speed up the rate of a reaction by over a million-fold, helping form stable products at a very fast rate and allowing information transfer to occur more efficiently.

An enzyme is the catalyst of a biological system. Enzymes are usually made of either amino acids or RNA and often with the help of a metal ion or two. Some of the first enzymes were most likely ribozymes (enzymes made of RNA), although most of the enzymes in current organisms are proteins. Ribozymes have certain abilities that make them much more likely precursors to life. For one, some ribozymes may self-catalyze—that is, a specific type of foldable RNA strand actually clones a part of its sequence without the aid of other large molecules.[18] Ribozymes are also capable of catalyzing reactions such as adding on more nucleotide bases to other RNA strands. In doing so, they are adding complexity and information to other RNA molecules.[19] And most importantly, certain ribozymes are capable of binding amino acids (the building blocks of protein) together, allowing them to form the catalysts most commonly used in organisms today: protein enzymes.[20]

To understand the usefulness of a catalytic reaction, imagine a robot in a factory, helping to put pieces together quickly to assemble a product. A faster, more efficient robot is like a catalyst because it helps the assembly line produce more of whatever it makes. Now imagine that instead of making car parts, for example, the assembly-line robots were making copies of themselves: more assembly-line robots. When this type of robot successfully builds its clone, there are now more robots working around the clock to build copies of themselves. As more and more robots are completed, the number of robots increases exponentially. The rate of production goes up as the production itself occurs.

The situation above, transplanted from the assembly line into chemistry, is known as an *autocatalytic reaction*. Autocatalysis is a reaction whose product is the catalyst of the same reaction that made it. As mentioned earlier, ribozymes are capable of cloning themselves in such a manner, as well as creating new RNA molecules and long protein chains (which can also function as enzymes).

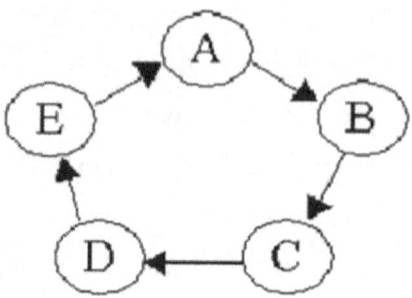

Figure 6.8 Autocatalytic set

In most organisms alive today, proteins help catalyze the formation of RNA, which suggests that ribozymes catalyzed the creation of other enzymes that could increase their own production and make it more efficient. A catalyst (RNA) made a catalyst (protein) that could help make the first catalyst again (RNA). This type of loop is known as an *autocatalytic set*, where instead of creating another copy of itself, a catalyst produces the first link in a chain of reactions that eventually lead to a new copy of the original. Catalyst A helps form Catalyst B, which in turn helps to create Catalyst C. This chain of information continues, until Catalyst A is created again at the end of the autocatalytic set. As long as the molecular resources are available, the autocatalytic cycle repeats, and the molecular messages continue to cycle around. This self-reinforcing cycle is known as a positive feedback loop.

Catalysts are so widespread and helpful to the functioning of molecular systems that they undoubtedly played an important role in the emergence of complex organic matter. Stuart Kauffman, a theoretical biologist and complexity researcher, proposed that autocatalytic sets might have directly given rise to the first organic life forms.[21]

An autocatalytic set isn't exactly "life" as we understand it today. It does consume the building-block molecules around it ("food") to create more copies of itself ("reproduction"), and it does grow. It possesses a number of characteristics of what we define as "life" and is the beginning of a chemical communication network, so it is a good starting point. However, life as we understand it today contains the much more stable DNA as the information-replicating molecule, rather than RNA.

6.9 Why DNA?

After years of selection and various chemical reactions, DNA was the molecule that outcompetes the rest for carrying the genetic blueprint of a cell. At first, this may seem surprising—after all, we already mentioned that RNA is a molecule capable of self-replication and catalysis, and it is postulated to have arisen as genetic material before DNA. Yet it is DNA, rather than RNA, that is the genetic blueprint found in bacteria, archaea, and eukaryotes. This seems surprising, until one looks closer at the chemical structure of DNA, which indicates much more stability, making it a better candidate for the transfer of information to the next generation.

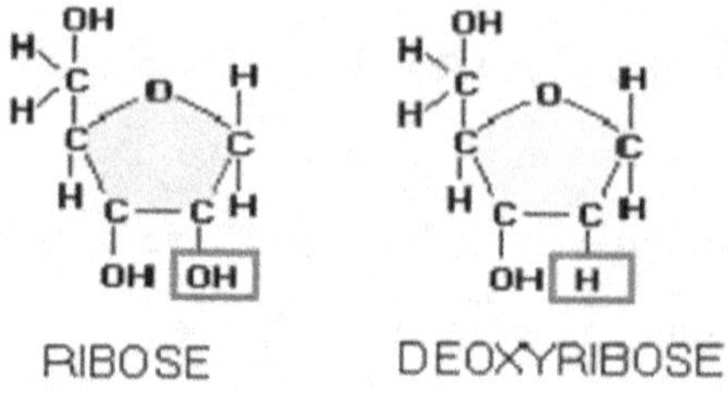

Figure 6.9 Ribose (RNA) vs. deoxyribose (DNA), both in ring form

DNA and RNA are both made up of three parts: a nucleotide base, a sugar, and a phosphate backbone. Although the phosphate backbone and three of the nucleotide bases are the same for both molecules, the differences in the sugars and the final base pair lead to a substantial difference in stability. DNA, short for "deoxyribonucleic acid," and RNA, short for "ribonucleic acid," contain "nucleic acid" (a nucleotide base) and a "ribose" (sugar). The difference between the two is a missing oxygen atom in DNA, normally bound to carbon as hydroxide (OH) in RNA (*Figure 6.9*).

The change seems minor, but the difference between a hydroxide (in RNA) and a hydrogen (in DNA) has an important effect on the stability of the molecule. On average, a C-O bond is 55 kJ/mol weaker than a C-H bond, meaning that the C-O bond is easier to break.

Because it is easier to break, the C-O bond in RNA can go through various chemical reactions, such as detaching from the phosphate backbone and breaking apart the RNA molecule. This cleavage of the backbone can occur when RNA is exposed to metal ions or if it folds in a specific manner due to environmental conditions (like temperature or salt content).

The deoxyribose in DNA is more stable because the C-H bond takes more energy to break and is less chemically reactive, making a deoxyribose backbone more stable than a ribose backbone and better at retaining its information content.

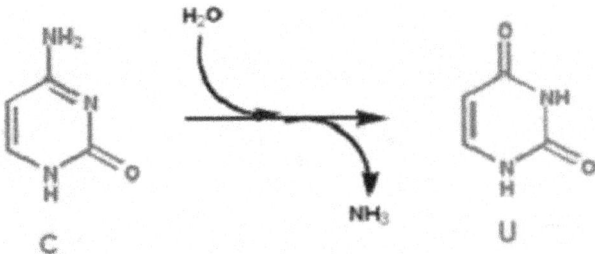

Figure 6.10 Cytosine chemically turning into uracil by using water (H$_2$O) and releasing ammonia (NH$_3$)

Another difference is in the nucleotide base-pairs. These are molecules that make up the genetic code: information that when translated into proteins, determines almost all characteristics of an organism. There are five nucleotide bases: adenine (A), guanine (G), cytosine (C), thymine (T), and uracil (U). A, G, and C are common to both DNA and RNA; T is found only in DNA, and U is found only in RNA.

111

These bases contain information regarding which proteins are to be formed from the sequence. As such, a mistake in the base pairs could prove detrimental. To prevent this, there are mechanisms in place to try to repair such mistakes. C can spontaneously decompose into U, changing a message in the genome vital for the cell's survival. But in RNA, because U is used rather than T, it would be very difficult for an enzyme to recognize if the U needs to be repaired into a C or if the U is naturally supposed to be there. Because DNA uses T instead of U, all repair enzymes recognize U's as mistakes and repair them to C's, conserving the messages necessary for survival within the DNA.

Base pairs are also important to the double-stranded nature of DNA and RNA. C and G can "pair up" by having three sets of atomic groups form hydrogen bonds with three sets of atomic groups across from them (in the other strand). A hydrogen bond is an interaction between a molecule with a bigger positive pull (like hydrogen) and a molecule with a strong negative pull (like oxygen or nitrogen). The difference in charge causes the two sets of molecules to attract each other. Compared with chemical bonds, the hydrogen bond is relatively weak, around 5-30 kJ/mol, while atomic bonds can be anywhere from 100-1,000 kJ/mol.[22] It is like comparing the pull of a refrigerator magnet to the pull of a rare earth magnet—you can pull a fridge magnet off the door easily, but even a team of horses couldn't pull anything from a rare earth magnet.

Although hydrogen bonding is a weak interaction, the enormous number of hydrogen bonds involved combines to provide a surprising amount of stability. If left single-stranded, RNA and DNA make most of their hydrogen bonds with water molecules in the cell's environment, forcefully "hugging" the water molecules with their hydrogen bonds. This interaction brings order to the system, which is unfavorable for small, simple molecules, like water, which would easily break off from RNA or DNA and mix freely with the surrounding elements. When a base pair binds with hydrogen bonds, it displaces the water molecules, allowing water to float freely once more, increasing disorder in the surroundings, and increasing stability in the DNA or RNA molecule. This process continues like a zipper closing, and complementary DNA or RNA bind to each other via hydrogen bonds.[23]

So, forming a double helix is more energetically stable, and DNA helices have two advantages over RNA helices. First, although A can bind to either T (in DNA) or U (in RNA), the A-T bond is 9.8 kJ/mol more stable, meaning that a DNA double helix is almost 10 kJ more stable for every mol of A-T pairs.[24]

Second, the type of helices formed differs between DNA and RNA due to the different sugars used, so DNA forms a B-DNA helix formation, while RNA forms an A-DNA helix.

The B-helix transfers information more efficiently than an A-helix, due to the positions of the "grooves." A groove is an opening in the natural helix formation, where there is space between the two strands. There are major and minor grooves in all DNA helices. A major groove contains more information than a minor groove and contains the side of the base pairs with electrostatically different chemical groups. As such, an enzyme that needs to locate a certain base pair uses the major groove to find it. The minor groove does not contain such chemical groups, and the "lock" for different base pairs could be opened with the same "key," lowering the specificity, and thus the information output. B-helices formed in DNA have a large number of major grooves, big enough to fit a protein; A-helices forming RNA have small, narrow major grooves that are too small for proteins to fit in, lowering the information output from their base pairs compared to DNA.[25]

In short, as a molecule, DNA outperforms RNA in chemical and thermodynamic stability, as well as information conservation and transfer ability. These traits give DNA an advantage as a conservator of information, in most environments, causing it to be selected for as the reproductive transmitter of information to the next generation.

6.10 Self-Replication and Variation

What sort of environment would be ideal for a living, autocatalytic set to originate and maintain itself? Certainly not one where asteroid impacts are common. Certainly not one with lava flows, constant volcanism, and violent tectonic shifts.

114

However, that is exactly the hand Earth dealt to the first organic molecules. Life emerged during a tumultuous time. The first 1.2 billion years of the planet's history, which includes the origin of life, is known as the Hadean period. The word "Hadean" is derived from "Hades," the ancient Greek name for both the underworld for departed souls and the god who ruled it. Hades as a realm was not famous for its hospitality and neither was the environment of early Earth. Such an environment could have been easily more dangerous than a gently swirling pool of autocatalytic molecules could have handled. A makes B, B makes C, and C makes A; but what if C is destroyed by heat? What if B drifts away before it can react? Entropy would be the greatest threat to an emerging system, such as the first life on Earth.

How could early life have reproduced itself in such a chaotic environment? Any scenario, in which the reactants remained in the same space, would have provided a better setting for autocatalytic reactions to take off. The reactants would have needed an energetically stable home. This is where the lipid bilayer could have come into play. Perhaps the relative shelter of a protomembrane, emerging spontaneously from the chemical properties of lipids, could have prevented the necessary molecules from diffusing away from each other, and allowing them to be close enough to begin an autocatalytic set. From there, complexity could have bred further complexity.

Neurobiologist, Richard Pico describes the world of the protocell in terms of the constant creation and destruction of protocellular membrane sacs.[26] In this vision, each sac provides a short-lived protective microenvironment for chemical reactions to take place. Then, as each protomembrane breaks apart, succumbing to entropy, it dumps its newly created contents back into the surrounding medium. These contents mix and recombine, before being thrust into a new protocell: a new spontaneous lipid sphere.

We can think of these protocells as providing the newly formed molecules with a protective shield, preventing them from falling apart. They were safe havens for the creation of more complex molecules. Within these protocells, agents interact in close proximity to one another. Their interaction is frequent and unimpeded by the forces outside of the membrane. Because the agents within a protocell are free to communicate with one another through complementarity and are undisturbed, the stage is set for information networking and emergence. As the products of an autocatalytic set are consistently dumped back into the environment and then recombined within new protomembranes, they spread and take on different compositions. We can imagine the primordial soup becoming cloudier as larger molecules form. We would see different variations of self-replicating, growing molecules leading to molecular diversity. The foundations for a cellular metabolism could have been seen in the autocatalytic set's consumption of

116

smaller molecules ("food") that would have provided the fuel for its reactions.

Most importantly, as these reaction-cycles spread and varied, the basis for natural selection would have appeared. The autocatalytic sets that could most quickly synthesize molecules from the more abundant smaller molecules would have been best suited to spread in a prebiotic world, maintain their integrity, and finally replicate.

6.11 Evolution at the Level of DNA

We described how life began based on chemical evolution, through the selection of energetically stable molecules exchanging information with one another. We will now show how this process has led to energetic efficiency in bacteria, which used DNA as the genetic template for passing on information to their progeny. This evolution continued through the selection of more energetically efficient traits, by developing more effective information networks and the rhythmic entrainment of interactive autocatalytic sets, forming a metabolism. Evolution selects for cooperative chemical networks that are more energy efficient. In the fluctuations of energy availability in the environment, the forms that are selected and survive require less energy to evolve and persist. Their energy efficiency is facilitated by the network effect.

Evolution acts by selecting which organisms are best fit for a certain situation. In the cold, those that can avoid freezing, survive. When it floods, the creatures with the best swimming capabilities live on.

Genetic variability helps determine differences in survival strategies. Within any species, in any population, there are going to be differences in the DNA base pairs among individuals, which can lead to differences in the proteins that the DNA codes for. A difference in a protein changes the protein's function and alters the information it can transmit. These changes in function reflect on an organism's appearance or physical capabilities. DNA is the blueprint that codes for RNA and protein in all organisms; if a change is made to the DNA, the change is reflected in the products, which is reflected in the organism itself. This is similar to how a message is altered in a game of "telephone." DNA is like an instruction manual, with RNA reading the messages and using them to code for protein. If something is wrong in the instruction manual, the RNA doesn't always notice, and it can still make a protein, even if there's a mistake. In this sense, RNA is analogous to a speed reader. It acquires and translates the information, overlooking minor errors.

DNA changes between a parent organism and its offspring, in species which undergo sexual reproduction, can be primarily attributed to the intermixing of portions of DNA between two parents. A black mouse and a brown mouse would probably only lead to more black or brown mice. But where would a white mouse suddenly come from? How do new traits suddenly pop up in such a system?

The answer is mutation. Most observable mutations are harmful, but others are adaptive and selected to persist. Mutation is simply a change in a DNA sequence. It could be a one-base change that doesn't even change a protein, or a large addition or deletion of an entire DNA sequence. Any change is a mutation, and often it may lead to a negative result. But it can also lead to the introduction of new, useful, and energy-efficient traits into a species and, thus, genetic variability.

Mutations can be caused by a variety of factors at any point in an organism's life. Radiation and various chemicals can cause base pair changes, insertions, or deletions, depending on how they interact with a DNA molecule. Viruses can transplant their genetic code into another cell, changing its DNA's composition. There are even repeated sections of DNA, called transposons, that rarely, but naturally, copy themselves onto another segment of DNA. Most commonly, mistakes are made during DNA replication, where the replication enzymes can "slip" and add too many base pairs, skip some base pairs, or insert the incorrect base pair into a copy.

Imagine that DNA is like a written instruction for someone that is usually written as "don't yell." The messenger RNA will translate this message into protein, which will act on the command: don't yell. But, if a letter is changed, the message can be altered completely. It may turn into, "don't tell" or "don't sell." It will still work, but the protein will be performing the wrong function. If letters are removed, it can change the message to: "do yell," which means the protein will basically be doing the complete opposite. Even worse, the message could turn into "don't pell," which no longer makes any sense and causes the protein to lose function entirely.

In cell division of single-celled organisms, the daughter cell DNA contains practically all mutations of the parent's except those rare ones which are consistent with viability of the daughter cell. So we can say the daughter cells are clones of their parents. However, in multicellular organisms, the mutation is only passed on to the offspring, if the sex cells—cells that combine together to form an offspring—carry the mutation. Additionally, because sex cells combine with another set of chromosomes, there is a chance that the mutation, even if it were present in the portion taken from the mutated parent, could be hidden by the other parent's chromosome, though all the cells in the offspring's body will have that mutation in their genes silently.

Every change, large or small, can prove either advantageous or detrimental for an organism. It can be quite bad and can cause a protein to lose its function, like in the case of cystic fibrosis in humans. Those suffering from cystic fibrosis are born with a defect in their gene for the cystic fibrosis transmembrane conductance regulator (CFTR) protein on the cell membrane. This protein promotes the uptake of salt into cells—without it functioning properly, salt accumulates in the lungs and pancreas, causing the build-up of thick, sticky mucus, which causes breathing and digestive problems.

However, not all changes are detrimental. Some have little effect on the survivability of an organism, like the color of a mouse's fur. Black, brown, and white are very common colors for mice, and in very dark places, the color of their fur usually has no effect. However, if the environment changes and the mice need to survive in the woods during the day, the brown and black mice are more difficult to spot, so they have a selective advantage over the white mice, which are easier to spot and thus will have a higher probability of being eaten. If the area is covered in snow, however, the white mice would have a considerable advantage over the black and brown mice, and now the white mice would be selected for. This is Darwin's natural selection.

The organisms with the most advantageous traits for a certain scenario or environment are selected for, over time. Then, as more mutations accumulate in a population, new traits are selected for as the situations change. Part of a population can become separated by a river or a mountain range, causing environmental differences. Different traits are selected for, as different mutations continue to accumulate on either side, eventually leading to a completely distinct species. In this way, advantageous mutations continue to accumulate. This process is descriptive evolution, as described since Darwin.

6.12 Origins Through the Lens of Thermoinfocomplexity

We do not know with absolute certainty if life developed in the surface waters of earth or in the depths of the sea next to a hydrothermal fissure. It is likely, however, that the first proto-life was born from the self-organizing properties of simple molecules reacting in a network. Molecules in autocatalytic sets began to produce additional molecules. These mutually beneficial molecules developed cycles of production. This stochastic process repeated, as a response to the depletion of available Gibbs free energy in the environment, encouraging the development of more-energy-efficient organisms. Here, we see that the principles of Thermoinfocomplexity theory can explain how the complex macrostate of an organism emerged out of embedded and entrained energy within many information microstates. Natural selection acted on different forms—a panoply of reproducing, metabolizing, membrane bound sets—creating longer-lasting compounds and more complex living structures.

Competition, communication, cooperation, and selection worked hand-in-hand, at times limiting and at times amplifying the ordered complexity and structure of life forms in competition for the Gibbs free energy available in their environment. We may recall that in various complex structures, energy flow slows down by percolating through its complex information network. For the system to maintain its structure, a flow of Gibbs free energy is necessary. Systems that are better at capturing Gibbs free energy and using it more efficiently will be more likely to survive. This is the 'how' of evolutionary selection.

When looking at the origin of life through the lens of Thermoinfocomplexity, we can see life emerging and evolving by stochastic probability of persistence of energy efficient emergence of entities that result in progression of complexities from non-living to living matter. The same principles that sparked life in its most primitive form continue to shape even the most complex. Emergence begins with the stochastic process of interacting agents attracted to strange attractor paths, entrained cooperatively to create increasingly efficient complex patterns of being. This process starts with atoms, and leads to molecules and compounds, and from there reaches deep into every macrostate (with embedded microstates), from single cells to multicellular organisms, and finally to the collective behaviors of social insects and other animals. The process is scale-free and reiterated in stepwise emergence through the stochastic process of evolution that selects for complex adaptive systems that obey the laws of thermodynamics, as well as information and complexity theories. As we have noticed in this chapter, the major structural and functional components of the protocell—cell membrane chemical gradient, autocatalytic set, and self-replication—provided the foundation for the various information networks within the cell. Biological structures can be thought of as information networks, within the complexity of an adaptive system that helps the system to use less energy in the maintenance of its structure. Such systems follow the laws of thermodynamics, and when available energy is scarce, more complex information networked organisms are selected, as the available energy fluctuates. In this way, complexity builds on complexity, resulting in increasingly complex organisms and systems. The progressive interlocking of larger numbers of autocatalytic sets gives rise to the metabolic cycles and rhythms that make the more-complex organisms more-energy-efficient and robust. For this reason, as we shall see, the metabolic rates per gram of complex organisms are lower than those of simple ones.

In the next chapter, we will examine how simple information signaling shapes behavior in some of the simplest life forms on the planet: bacterial colonies and slime mold. We will see fascinating behaviors emerge with only a few basic molecular rules at work. We will follow how self-organization leads to the complex structures and behaviors of some of the strangest organisms, and how energy availability and scarcity affects form and complexity in the simplest organisms. We will also examine how energy efficient information transfer is selected for in surviving organisms. Our exploration will also give us a hint into how multicellularity may have emerged, and how complexity breeds energy efficiency and robustness.

Chapter 7 – The Rise of Biological Networks

7.1 Bacteria Rule the World Through Networks

7.2 Chemical Communication: Quorum Sensing

7.3 Network Complexity and Adaptability: Swim, Attack, Coordinate

7.4 Brainless Unicellular "Loners" or Multicellular "Slug"

7.5 The Messenger of the Slime Mold Information Network: CAMP

7.6 Toward Multicellularity When Famine Comes

7.7 Energetics Behind Emergence of Multicellular "Slug"

7.8 The Chemistry of Early Altruism

7.9 Self-Organization and the Power of the Fluctuating Chemical
    Gradient

7.10 The Evolution of Multicellularity

Figure 7.1 Diagram of quorum-sensing system

Figure 7.2 Branched colonial pattern formed by the *Paenibacillus*
          dendritiformis lubricating bacteria

Figure 7.3 Diagram of the life cycle of slime mold

Figure 7.4 Schematic diagram of the mechanism underlying
          control of CAMP secretion in slime mold

Figure 7.5 Spiral slime molds, spiral galaxy

7.1 Bacteria Rule the World Through Networks

Bacteria are a successful and ancient form of life. They are single-celled organisms with circular DNA and no internal, specialized compartments. Some say that bacteria rule the world. No matter where you go, you cannot escape them. Nor would you want to. Without bacteria, life as we know it would cease to exist. The typical human body has over 3.5 lbs. of coexisting bacteria and organisms living within it. As our heart weighs only 0.7 lbs. and brain just 3 lbs., our personal bacteria can be seen as one of the largest "organs" in our body. The healthy human body contains more than 100 to 300 trillion coexisting bacteria and yeast, compared to only 10 trillion human cells.[27]

Our internal (largely intestinal) bacteria do not act alone. They employ chemical communication to form complex structures along a hierarchical progression from 10^9 to 10^{12} bacteria. They form networks with more-efficient communication, using less energy per gram as their complexity increases. By acting jointly, they make use of any available source of energy in the environment, from deep inside the earth's crust to nuclear reactors. Under unpredictable and hostile environmental conditions, when the odds are against survival, the bacteria turn to a wide range of strategies for survival, such as frequent mutation and other adaptive collective responses.

We might ask: how do molecular networks at the single-celled organism level ultimately define the collective bacteria behaviors and their social interaction? How does the interaction among single-celled bacteria enable the spread of information and lead to dynamic population behaviors and complex biological networks?

The answer lies in the fact that there is an inherent degree of plasticity or flexibility in bacterial self-organization. The building blocks of the colony are themselves individual living organisms, each with internal degrees of freedom, stored information, and responses to external chemical messages. Each bacterium responds flexibly and even alters itself by means of modifying its genetic expression patterns.

At the same time, efficient adaptation of an entire colony to adverse growth conditions requires reorganization on all levels, which is achieved via effective signaling and communication between individual cells. Bacteria can communicate via a broad repertoire of biochemical agents, using chemicals as messages that can be received by other neighboring cells.[28] Such mechanisms are used to generate appropriate responses to environmental conditions. Biochemical messages are also used by bacteria in the exchange of information across colonies of different species and even other organisms. The colony behaves much like a loose arrangement of multicellular organisms or even a social community, with elevated complexity and plasticity that affords better adaptability to changing environmental conditions.[29] Bacterial signaling is thus an example of the emergence of biological complexity at an elementary level.

7.2 Chemical Communication: Quorum Sensing

To manage and maintain social cooperation in a large group of bacteria such as a biofilm, the bacteria need to send, receive, and recognize messages from colony members, as well as from other colonies of the same or other species, separating those messages from the noise of the surrounding crowd. They achieve this communication by using diverse chemical languages that utilize a wide range of signaling molecules called auto-inducers, such as acetylated homoserine lactone (AHL), oligopeptides, AI-2, γ butyrolactones, quinolones, amino acids, and many others.[30] Because these compounds are diffusible across colonies and because their accumulation correlates with cell density and species type, these compounds contain information about the metabolic context of the environment and the constituent species present. Bacteria are able to determine what species or food exist in their surrounding environment by sensing which chemicals are released by their neighbors. The use of this diverse assortment of molecules as a source of information, exemplifies the acuity with which bacteria sense, integrate, and modulate behaviors in response to ever-fluctuating environments.

In a process known as quorum sensing, groups of bacteria communicate with one another and coordinate their behavior and function together. Examples can be found in biofilm formation, bioluminescence, and sporulation. In principle, communication within the biofilm is similar to that of the cells of a multicellular organism. The accumulation of a stimulatory concentration of an auto-inducer can occur only when a sufficient number of cells, a "quorum," is present.

As an example, a typical bacterial quorum-sensing circuit is shown in *Figure 7.1*. In this type of system, the autoinducer is AHL, which is synthesized by a LuxI-type enzyme. The cytoplasmically synthesized auto-inducer diffuses passively through the bacterial membrane and accumulates both intracellularly and extracellularly in proportion to cell density. When the stimulatory concentration of the AHL is achieved, a LuxRtype protein binds to it and forms a LuxR-AHL complex. LuxR-AHL binds to the specific parts of the bacterial DNA that initiate gene expression, resulting in the translation of genetic information into working, functioning proteins.

The presence of these proteins leads to specific bacterial behaviors, forming the basis for the bacterium's response to this signal from its environment. Quorum sensing is part of the process of bacterial colonies forming biofilms, defending against predators, or searching for food together. The mechanism of quorum sensing is a good illustration of how complex adaptive systems function in the context of biological information networks. There is a constant two-way flow of information via chemical signaling between biological networks in the bacterial colony, and molecular networks within the bacteria cell itself.

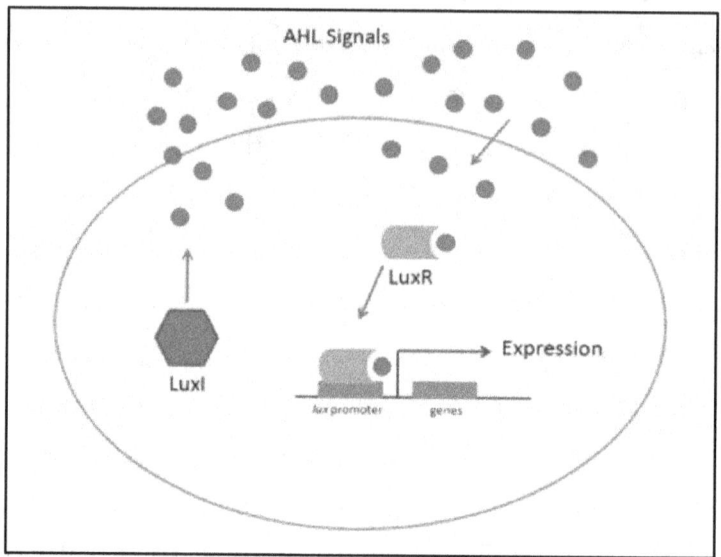

Figure 7.1 Diagram of quorum-sensing system.[31]

7.3 Network Complexity and Adaptability: Swim, Attack, Coordinate

To illustrate the ability of bacteria to respond to environmental constraints, we will explore the branching patterns exhibited by the *Paenibacillus dendritiformis* lubricating bacteria.[32][33] This class of bacteria has developed a special strategy to move across hard surfaces: they collectively excrete a special layer of chemical lubricant, upon which they can swim. As they swim, they push that chemical layer forward, paving themselves a new path as they go.

But a dilemma arises when the available food in the environment is not sufficient to sustain a dense population able to secrete enough lubricant for locomotion. When such conditions are mimicked in petri dishes in the lab using hard substrates and limited nutrients, new and interesting patterns are observed, as shown in *Figure 7.2.* When the bacteria receive information from the environment that indicates low nutrients, they collectively activate genes to emit chemicals called surfactants. Surfactants extract fluid from the available substrate in the petri dish, drying it to a crust, and then creating a lubricating layer with a well-defined envelope within which the bacteria can swim, using the fluid they absorbed. The task is not simple; the production of the lubricant requires collective action within a dense bacterial population to communicate food-depletion within the substrate. Such collective action is only possible with effective communication through quorum sensing. Some bacteria die just to release their content as food for the survivors.

Then, a new pulse of growth occurs through communication by quorum sensing molecules and the accumulation of the food supply. The entrainment of all the attractor paths formed by rhythmic food supply and growth, leads to the self-similar fractal patterns we see in *Figure 7.2.* These complex branches emerge due to the individual interactions of the cells.

The first cell to sense "hunger" begins to release a quorum sensing molecule, and because there are not enough resources to make this a continuous process, it releases that molecule in bursts. This molecule attracts the other cells, which begin to move toward the initiating bacterium. When these cells receive the molecule, they release their own signaling molecules in response, creating a positive feedback loop. Because membrane molecules must "reset" after interacting with a quorum molecule, the different bacteria cells end up synchronizing with one another, based on their location in synchronous pulsations. The pauses between interactions allow for the formation of intricate repeating patterns driven by oscillations in the amount of Gibbs free energy (food) available.

The colony carefully adjusts the lubricant viscosity and its overall production rate to generate specific branch structures with precise widths, according to the hardness of the substrate and the level of food. This strategy works well for lubricating bacteria, as their chemical communication within the network solves the problem between two constraints: the high bacterial density necessary for collective movement, and a level of food that is insufficient to support high bacterial densities. Here we see a clear example of dynamic information signaling and the emergence of complexity, manifested in the fractal patterns of the microstate and macrostate of bacterial growth.

Figure 7.2 Branched colonial pattern formed by the *Paenibacillus dendritiformis* lubricating bacteria, reminiscent of organized complexity (Chapter 1) of snowflakes but in reality a result of repeated bit strings, as denoted by Murray Gell-Mann as effective complexity.[34]

In nature, many bacteria break down complex organic polymers, requiring the concerted action of many cells. For example, the predatory Myxobacteria utilize a "wolf pack" strategy to attack and digest their prey organisms together. They release digestive, extracellular enzymes to break down the structure of the prey, before it can be consumed.[35] Interestingly, groups of *Mycobacteria xanthus* cells can migrate in relation to chemical agents, but individual cells cannot.[36] The bacteria require a threshold level signal to begin movement. *M. xanthus* grazes on cyanobacteria in ponds. However, the aqueous environment can dilute both the digestive enzymes and the liberated nutrients, impeding the bacterium from catching the chemicals on which it feeds. Thus, the predators construct spherical colonies and trap prey organisms in pockets where their enzymes can be more concentrated, bursting the prey cell and keeping the released nutrients in one spot. In this way, the act of predation becomes much more efficient.[37]

This strategy is the result of a buildup of chemical complexity within the dense cellular information network. Information exchange between cells of form the network, allows for better hunting. This process comes about through the stochastic process leading to selection of the most effective information network.

In another example, the phytopathogen *Erwinia carotovora* releases exoenzymes into the environment that can degrade plant cell walls. This process occurs under the control of an AHL quorum-sensing system (see *Figure 7.1*). The process ensures that the bacteria will only invest their cellular capital in digestive exoenzyme production once there is sufficient bacterial cell density to effectively attack the plant structure. The bacteria are prevented from wasting energy resources when a quorum does not exist to produce an effective concentration of exoenzymes to digest the plant's cell walls. Bacteria, a successful and ancient form of life, benefit by living socially and forming complex information networks.

7.4 Brainless Unicellular "Loners" or Multicellular "Slug"

Slime mold is a tiny fungus-like organism that feed on microorganisms living on rotting logs and other decaying surfaces. Slime molds cooperate to accomplish something that would be impossible alone—a cooperative form of survival. When we look at the biological world, we would be hard pressed to find a more elegant example of emergent complexity than that of cellular slime molds. These organisms thrive all over the world in a wide variety of climates and ecosystems. Different varieties take on many different colors, from Wolf's Milk, to Bubble Gum and Red Raspberry, and on to Chocolate Tuba and Tapioca.[38] Across the species, many take on a gelatinous appearance.

For over a century, the slime mold has captured the attention of biologists everywhere. Despite a simple metabolism and a limited number of genes, slime molds behave in ways often thought to be exclusive to more complex organisms. Their popularity extends beyond the realm of mere scientific inquiry. Representations of slime molds appear in Hieronymus Bosch's famous 16th century triptych, *The Garden of Earthly Delights*.[39] No representation of nature would be complete without them.

D. discoideum amazes scientists for one primary reason: depending on the environmental conditions, *D. discoideum* cells act either as individual, unicellular "loners," or together as a social, multicellular "slug." *D. discoideum's* capacity to switch organizational strategies in response to informational cues from its environment has warranted a vast amount of attention in the scientific community. When these slime mold cells are in a nutritious environment with abundant Gibbs free energy, they live as separate individuals. With abundant resources, they act as unicellular organisms. However, when their bacterial sustenance approaches depletion, the unicellular actors adopt a qualitatively different behavior; they huddle together to form a cooperative community, a rudimentary multicellular organism.

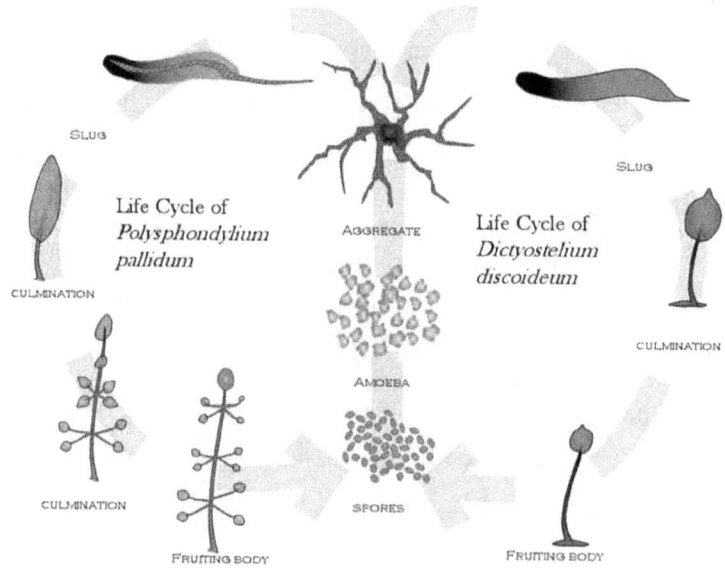

Figure 7.3 Diagram of the life cycle of slime mold [40]

Slime molds do not have a nervous system, much less a brain. They are single cells. They cannot, in any conceivable way, "think" about whether to join together or to live apart.

138

Furthermore, they have no "leaders" to make any sort of decisions regarding their collective fate. But this amazing behavior does not happen by chance either. This case study is crucial to our understanding of communication as the root of cooperation, as well as energy efficiency as the root of the emergence of complex forms in the natural world. It will also reveal some of the advantages and disadvantages of cellular communal living that leads to multicellularity.

7.5 The Messenger of the Slime Mold Information Network: cAMP

As we mentioned earlier, the cells of *D. discoideum* live independently when their environment has an abundant food source, such as bacteria. As food diminishes and individual cells begin to die of starvation, they secrete a chemical called cyclic adenosine monophosphate, or cAMP. The starvation starts a chain reaction that releases more cAMP from the starving *D. discoideum* into the environment. It is a positive feedback loop. Within a community of individual slime mold cells, the cAMP emission of one cell results in a chemical cascade involving all the *D. discoideum* cells in the area. The rapidly increasing concentration of cAMP among the dead or dying cell population results in the formation of a collective mass of cells.

Through a series of chemical reactions within the cell, cAMP ultimately makes each cell extend a pseudopod. In this way, cAMP is a chemical signal for communicating the extension of a pseudopod. cAMP contains the information responsible for a chemical cascade that brings about the change in morphology and behavior. The pseudopod acts as a temporary projection allowing for cell movement. The cell extends its newly formed pseudopod, and moves toward the source of the cAMP, in a process called chemotaxis (movement toward a chemical signal). Areas of high cAMP concentration attract all the nearby cells, which emit their own cAMP if they are also starving, further strengthening the signal. It is a classic case of positive feedback, leading to the emergence of the collective slime mold, or multicellular "slug." Each individual follows basic rules: a cell produces cAMP when it is hungry, and it moves towards the cAMP put out by other starving slime mold individuals.

Although this information transfer is not optimal, cAMP is nonetheless a chemical messenger whose presence induces a collection of individual cells to move toward each other by means of the information transfer—the first step toward building a coherent whole. As a chemical messenger, cAMP is very slow compared with a neuron, which communicates a signal over a much greater distance almost instantaneously through electrical signaling. cAMP diffusion across a typical slime mold substrate, such as a forest floor, is even slower than hormonal diffusion, which occurs in our human bodies whenever chemicals free-float their way through our bloodstream. In either case, chemicals relay information via diffusion. Nerves in animals of a higher evolutionary complexity communicate faster than both, with especially efficient electrical depolarization among neurons within the network.

As these biological communication networks grow more interconnected and electrochemically efficient, such as within a human body, information transfer quickens and contributes to the more-efficient use of energy.

If the human body lacked nerves and relied instead on the diffusion of hormonal messages from our hand to our brain, we would be unable to react with the speed necessary to avoid harmful situations, such as the act of placing our hand on a hot stove. It could take seconds to minutes before we would be able to react, and by then the damage would be irreversible. Before the evolution of the nervous system, simple organisms like the slime mold had to rely on nothing more than diffusion and simple proximity cascades, which occurred between adjacent cells over relatively short distances. Nonetheless, cAMP serves as a clear message when energy in the environment is low, allowing for survival through sharing food energy among the survivors.

7.6 Toward Multicellularity When Famine Comes

In addition to slime molds, bacterial cells also secrete cAMP. Bacteria emit cAMP for the same reason the slime mold does: to indicate the scarcity of food resources. The close relationship between cellular slime molds and these types of bacteria suggests that they share this cAMP signature from a common ancestor. But what began as a chemical byproduct of starvation was co-opted by predators to identify prey bacteria. In this case, the interpretation of information held by the cAMP signal is completely different than it is when cAMP is used for aggregation. The information transfer occurs between predator and prey, allowing the slime mold to locate its bacterial food. What is unique to *D. discoideum*, is that it also utilizes environmental cAMP to aggregate into a slug. By tracing a bit of the diversity of cAMP use among the slime mold and its cousins, we begin to get a picture of how such a complex behavior and its resulting, emergent structure may have evolved from something originally quite simple. Even the world of slime molds has a rich history.

It is important to note that a critical mass of cAMP is required before the development of the slug really takes off. Systems have a tendency to maintain the status quo until the provision of a critical amount of a specific signal. Once this critical level is reached, a change can occur that results in emergent complexity within the system. *D. discoideum* cells are no different. A number of feedback mechanisms involving protein receptors in the cell membrane work to dampen a weak cAMP signal. In this way, a few hungry outlier cells will not start a cAMP cascade if most of their neighbors still have abundant bacteria to eat. This homeostasis is maintained until a critical mass of hunger and subsequent cAMP production begins an irreversible cascade of chemical signaling. At this point, the cAMP waves build and build, resulting in the aggregation of individual slime mold cells into a "slug."

As with any other complex adaptive system, positive and negative feedback loops play a major role in the self-organization and self-regulation of slime molds. As illustrated in *Figure 7.4*, increased concentration of extracellular cAMP causes increased production of intracellular cAMP, which may then leave the cell and trigger this reaction in neighboring cells. In this manner, cAMP is produced by a process of positive feedback, where cAMP stimulates its own production. However, if this mechanism were to continue indefinitely, cAMP would quickly grow beyond reasonable bounds. There must be another mechanism to limit its production beyond a certain level. In 1987, Martiel and Goldbeter proposed a model for the cAMP receptor and showed how the dynamics of the receptor itself could result in an oscillatory secretion of cAMP.[41] When the receptor is in an active state, meaning cAMP is molecularly bound to the receptor, it can combine with and activate adenylate cyclase. Adenylate cyclase is an enzyme that catalyzes the formation of cAMP from ATP and increases the rate of intracellular cAMP production. Hence, as the extracellular concentration of cAMP increases, the rate of intracellular cAMP production also increases. Under specific circumstances, increased intracellular cAMP concentration leads to increased extracellular cAMP, which inactivates this receptor. As the

143

fraction of active receptors decreases, the rate of cAMP production also decreases, leading to very low cAMP levels. This negative feedback step oscillates with the positive feedback loop.[42] The entrainment of a series of biochemical cycles, throughout the attractor pathways of the constituting molecules, results in the emergent, energy efficient cascade. We will provide experimental evidence for this in Section 7.8 of the following pages.

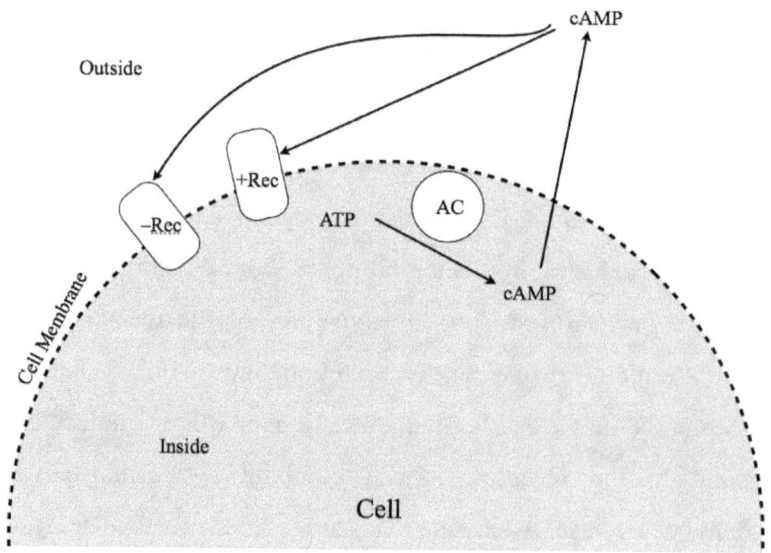

Figure 7.4 Diagram of the process of cyclic cAMP secretion in slime mold: cAMP binds to its receptor (+Rec) thus activating the enzyme adenylate cyclase (AC) resulting in production of further cAMP pushing the transport of cAMP to the outside of the cell. Here the positive feedback loop is shown by "+" and the negative feedback loop is indicated by "−". Extracellular cAMP inactivates the cAMP receptor of the membrane by

144

converting it to (-Rec) thus inhibiting further production of cAMP within the cell.

The communication between individual cells leads to surprising emergent results for the aggregate "slug." The emergence of the multicellular "slug" carries a selective advantage for the population as a whole. Once the cells communicate and collaborate within this network, they are able to reproduce even when local resources have been depleted. From out of the slug, a "stalk" begins to form, composed from the bodies of dead cells. The stalk grows in the direction of the soil surface, and ultimately reaches into the open air above the ground as the starving slime mold cells search for new food sources, climbing on top of one another, and then dying. At the top of the stalk are fruiting bodies, spores of *D. discoideum* protected by strong cell walls. The spores poke out of the ground until another organism, such as a worm, insect, or mouse comes by and picks them up, carrying the spores to greener pastures where the slime mold cells are able to thrive. Once the spores receive a chemical signal that they have reached a new, nutrient-rich destination, they begin to feed and multiply, continuing the life cycle of their species. The result is a new generation of individual, dispersed slime mold cells. As long as there is plenty of food around, those individuals will remain as dispersed, single-celled organisms.

7.7 Energetics Behind Emergence of Multicellular "Slug"

As we mentioned earlier, Thermoinfocomplexity theory describes the dynamic interplay of energy and information, and the stochastic evolution of complex adaptive systems. According to this theory, biological systems use information through signaling and communication to improve their energy efficiency. When the Gibbs free energy in the environment is scarce, evolutionary processes favor those systems that need less energy to survive and reproduce. In the case of slime mold, when the food supply is low, the single cells send and receive signals that enable them to come together to create a multicellular "slug."

In a study conducted by Wright and Bloom, slime molds at several developmental stages were incubated with radioactive glucose, and the amount of radiochemical CO_2 yield was determined.[43] [44] The results indicated that with the progressive differentiation of slime mold, there was a tenfold decrease in the relative radiochemical yields of CO_2, indicating a lower metabolic rate per cell when slime mold cells form a biological network. In other words, within a group, the slime mold expends less energy per cell.

When slime mold cells come together and form a multicellular organism, they have limited food options, and at least in part, receive necessary energy by utilizing their internal reserves: breaking down internal cell proteins and metabolic cycle intermediates, in the process of autophagy: the cannibalism of weaker neighboring cells. The new emergent "slug" of slime mold cells is more energy efficient, because it shortcuts its metabolic pathway by means of self-ingestion and collective hunting through cAMP signaling. In this way, information exchange and complexity contribute to greater levels of energy efficiency within the emergent organism.

7.8 The Chemistry of Early "Altruism"

The slime mold "slug" undergoes a process of reproduction, through the formation of spores, which act as an escape pod for *D. discoideum's* genetic material. The cells left behind—the ones that formed the spore's stalk—die of starvation, whereas the spores go on to establish new colonies of slime mold in places where resources are more abundant. This behavior has been described as "altruism." Many individual cells sacrifice themselves in the interest of their progeny within the fruiting body, which passes on their genes to the next generation. Though some of its members survive, we are still left with a clear case of individuals behaving in a self-sacrificing manner, for the good of the multicellular "slug" rather than the individual.

cAMP is part of the chemical family known as acrasins, which are found throughout the slime mold lineage in various forms. They are all related to a slime mold's aggregation into its multicellular blob form, causing the transformation of a group of individuals into a single entity. The word acrasin was chosen to describe cAMP after Acrasia in Edmund Spenser's sixteenth century epic poem, *The Faerie Queene*. In the poem, Acrasia is notorious for seducing men and transforming them into beasts. She takes away their individual wills, much like her chemical namesake: the acrasin group of chemical messengers. The name Acrasia was inspired in turn by the Greek word akrasia: lacking command over oneself. Small amounts of acrasin cause cellular slime molds to unwittingly act against their own interest, building stalks at their own expense. And none of the slime mold cells may disobey the chemical signal, as it is encoded within their DNA.

During times of scarcity, the modern slime mold is awash in cAMP. The cells of these slime molds are unable to distinguish between the cAMP emitted by the bacteria on which they feed, and the cAMP coming from a hungry neighbor of the same species. What appears to result is an altruistic action initiated by individuals that are not particularly altruistic by nature, but rather are expressing their individual hunger. Here we see that unconscious self-interest through chemical signaling and communication can lead to what appears to be cooperative or altruistic group behavior in the service of the larger population and at a cost to the individual actors. In other words, slime molds communicate within and between individual cells, and show what appears to be cooperation to us observers, leading to the survival of their species. The most efficient communicators of information compose the individuals within the fruiting body that is selected to pass on the slime mold's genes through the spore. This whole process has resulted from Gibbs free energy scarcity and the emergence of a more energy efficient aggregation of communicating cells. The dead cells of the stock have died in search of food while other cells have climbed up the stock to result in fruiting body. It seems a stretch to call the stock cells altruistic.

7.9 Self-Organization and the Power of the Fluctuating Chemical Gradient

cAMP has another effect on a population of individual slime mold cells. Just as patterns emerge in the inorganic world in snowflakes and hurricanes, the waves of cyclic AMP that pulse through a slime mold population create intricate, emergent patterns of their own. Incidentally, the interaction of cAMP concentrations in the formation of the multicellular "slug" creates overlaying spiral patterns. The spirals emerge naturally without the direction of a single individual cell. The aggregation of slime molds reminds one of the gentle spirals of a galaxy. Although the physical properties of the two systems are quite different, both are testaments to the beauty of emergent patterns following the path of pulsating energy flow through the strange attractor paths.

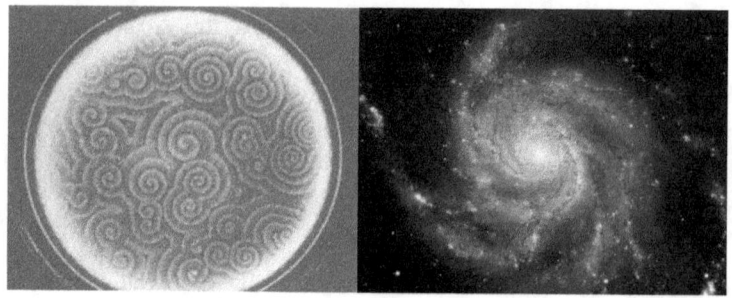

Figure 7.5 Thousands of slime mold cells in a petri dish, arranged in a spiral pattern, spiral galaxy.[45] [46]

And while the result looks a bit like the concentric patterns raked in the sands of Japanese rock gardens, the spiral slime mold pattern is simply another case of elegant organization emerging from the interaction of individual units following a handful of basic natural laws.

The key to these beautiful concentration gradients comes from the process of self-oscillation, mentioned above. The amount of cAMP emitted by any particular cell is determined by the amount of cAMP it encounters in the local environment. This output represents a nonlinear function. In this case, the output of any particular cell has an impact on the input of all its neighboring cells. The neighbors' input is then changed, and their own cAMP output will be different. The result causes the cyclic AMP output of each cell to fluctuate up and down, depending on the oscillations of key chemical supplies. This supply is itself linked to oscillations of available Gibbs free energy. The entrained oscillations of each cell responding to its environment create the global spiral pattern of cAMP concentration that we observe. But it does not end there. *D. discoideum* cells move toward the areas of highest cAMP concentration. If there are spiraling concentrations of cAMP, we should expect to see the movement of cells following a comparable pattern. This combined pattern of movement is known as the entrainment of strange attractor paths. In the case of the slime mold, entrainment leads to the beautiful spiral patterns that we see in cells moving via chemotaxis toward the center of the spiral.

Notice the power of the chemical gradient. Entropy can be defined as the lack of extractable information. Life exists because it is able, at least temporarily, to slow the dissipation of energy flow, by capturing it within information networks slowing down entropy. When the slime molds establish a chemical gradient, it creates conditions capable of reducing entropy. Systems of adaptive organized complexity will always display a gradient, allowing for the transmission of information beneficial to the persistence of that system. The cell membrane, so integral to the emergence of the first life on Earth, is perhaps the most fundamental divide between external and internal gradients. The cells of an individual slime mold extend gradients beyond the boundaries of their own membranes. In all such gradient systems, information appears in regular patterns, even if those patterns are created—as in the slime mold—from the bottom up. In the slime mold, the pattern is as simple as a concentration of a single chemical across a small portion of the forest floor. The level of concentration lays the blueprint for a spatial emergence, creating a single fruiting "slug" with symmetry, as well as an internal network of communication, where before, there was nothing more than lone cells that used energy less efficiently.

This elegant, mathematical formation is a textbook example of self-organization. Starving cells pulse waves of chemicals into the environment, culminating in an inward spiraling motion to form a multicellular "slug." Imagine each bit of information as a tiny point of light. An intricate, real-time tapestry of form develops as the vibrant information network comes into existence. Thus, when single-celled strategies fail, a multicellular slug emerges that can accomplish what the individual cells could not: the formation of a fruiting body to launch individual organisms towards greener pastures. The genetic information in a population that behaves this way survives the downsides of energy availability in its environment through communication between individuals. In the end, the slime mold multicellular organism is more energy efficient, and it out-competes populations that do not develop cooperative communication networks. It has taken billions and billions of generations of selection for this level of communication and organization to arise and endure. It does so by following surprisingly simple rules. The emergence of complex life arises at each step, becoming shorter and shorter as it builds upon the previous one.

7.10 The Evolution of Multicellularity

When individual slime mold cells huddle together, they face new challenges in addition to receiving adaptive benefits. E.O. Wilson points out that in laboratory tests, individual slime mold cells do better on their own than they do as a group, so long as they are provided with an abundance of food.[47] Huddling together is a response to food scarcity, and the species has better chances of surviving through tough times in a network of communication. Once individuals come together, many may be trampled, and some may even be cannibalized by their new neighbors. Additionally, when a group of organisms huddles close together, they need a mechanism that allows each organism to get enough energy to survive. Food and information outside the huddle will have to diffuse across a certain distance to reach each cell. Communication networks have evolved and increased the size and number of huddling organisms, whether they are individual cells, plants, or animals. Communication networks are the root conditions for multicellularity.

The slime mold communication network is chemical, and it depends on cAMP. However, communication need not be carried out by cAMP alone. Hormones and pheromones are chemical signals that facilitate chemical communication both within and among larger animals. Other multicellular organisms employ extremely complex circulatory systems that ensure all of their cells receive enough energy. At a much more basic level, the polarization and depolarization of the cell membrane, brought about by the sodium-potassium pump discussed before, is a ubiquitous cell membrane transport mechanism. This mechanism allows for the selective permeability of the lipid-bilayer cell membrane. It gives the cell membrane the ability to take in certain molecules, such as potassium, and to eject others, such as sodium. Here, at the most elementary level, we see the mechanism of the emergence of information transfer, setting the stage for the evolution of more-complex communication and information networks.

A group of cells will often increase in size and density first, in a stochastic process, before the organisms with improved communication networks, known as super-communicators, dominate the niche. Multicellularity has evolved on at least thirteen separate occasions in the history of life.[48] John Tyler Bonner may have suggested that larger cell masses were the foundation from which the new arrangement, i.e., multicellular organisms evolved based on the efficiency of their information communication. The new networks evolved as a way to keep up with the increasing demands of the diffusion and distribution of energy. Given the new needs of a cluster of cells, multicellularity was adaptive. It eased the flow of energy and decreased the metabolic rate per gram of the group as a whole, through entrained biochemical attractor pathways. This adaptation served as a selective advantage for survival when less energy was available. For this reason, multicellularity evolved independently in various instances, across many taxa. Fungi likely evolved multicellularity more than once. Brown algae began multicellularity when the rigid cell walls failed to separate upon division as a result of a mutation. The ciliate known as Sorogena is unrelated to the slime mold, but it nevertheless employs a very similar reproductive strategy. When resources are scarce, separate cells come together to form a stalk that rises upwards with a fruiting body. Multicellularity evolved among various taxa, including the red algae, the foraminifera, the radiolaria, the archaebacteria, blue-green algae—the list goes on. In every case, cells and aggregates coalesced in search of food: a source of Gibbs free energy. As they grew larger, only those with more-energy-efficient networks were able to survive.

In this sense, the story of the slime mold is a microcosm of the emergence of multicellularity itself. The story of multicellularity, in turn, is part of a greater pattern of the huddling of large numbers of entrained, oscillating strange attractor pathways of biochemical processes that we see as communication and cooperation. In the previous chapter, we looked at how life emerged from non-life. Its emergence depended on the stability of temporary membrane sanctuaries, where chemicals could come together and interact, exchanging information as they arrive at a lower, more stable, energy state. Early life needed a protective cocoon, in the form of a membrane to allow for the proliferation of complexity. Living things had to take in certain molecules, organize them, and expel others. The individual molecules were often destroyed or changed in the process of the emergence of new complex forms, but nevertheless the whole went on to survive and reproduce using information transfer to reach the most energy efficient, stable state. The whole achieved long-term stability, allowing it to evolve. With slime molds, we have seen a similar phenomenon, at a different level of complexity. The agents, in this case, are individual slime molds, rather than individual molecules. Their communication is accomplished via cAMP and other signals, resulting in the selection of the most energy efficient forms. Complex information transfer networks are more efficient, and are able to survive harsh environmental conditions, such as the scarcity of Gibbs free energy. The result—stabilized over countless generations by natural selection—is a sustainable, self-propagating system that adapts to the ups and downs of energy availability in the environment.

The examples described in this chapter illustrate how complexity arises through communication within biological information networks, resulting from the overall efficiency of the network and leading to the evolutionary success of more complex systems. Perhaps it is useful here to recall that all the chemical communication described for bacteria and slime molds is rooted in the stochastic retardation of energy flow, starting from photons to atoms, to molecules and living organisms. The foundation of the hierarchy of complex forms is based on the oscillation of Gibbs free energy through the whirlpools of information microstates, which creates the fractal basis of emergent macrostates. The slime mold multicellular "slug" is a complete example of this process. The next chapter describes the progression of this process that leads to the emergence of truly multicellular organisms, including embryos and frogs.

Chapter 8 – The Rise of Complex Organisms

Figure 8.1 Diagram of the cell with labeled endoplasmic reticulum, lysosomes, mitochondria and other organelles

Figure 8.2 Relationship between whole organism metabolic rate and body mass for prokaryotes, protists, and metazoans

8.1 Cooperation Within Our Cells: Mitochondria and Chloroplasts

Our bodies are composed of anywhere from 10 to 100 trillion human cells, 1,000 trillion bacteria, and 60,000 trillion

(60 quadrillion) symbiotic mitochondria.[49] [50] [51] [52] These are our symbiotic parasites that have their origin from our cells' devouring other bacteria and using them in a symbiosis of more energy efficient cells. So our guest mitochondria live within our cells cooperatively. They slow down the rate of energy dissipation within our body. We are a superorganism of multiple cooperative agents.

Within the flattened membrane of the endoplasmic reticulum, which winds through the cell, connected to the nucleus, the cell's proteins are packaged and transported to different parts of the cell. To accomplish this task, the organelle places specific "labels" on each of the molecules it creates. Another organelle, the lysosomes crunch-up waste products by trapping them in a membrane pocket known as a vacuole and then expelling the waste. Other organelles, the mitochondria— around 600 of them in each cell—employ ion pumps to manufacture adenosine triphosphate, ATP. Every day our bodies recycle a mass of ATP equal to our entire body weight.[53]

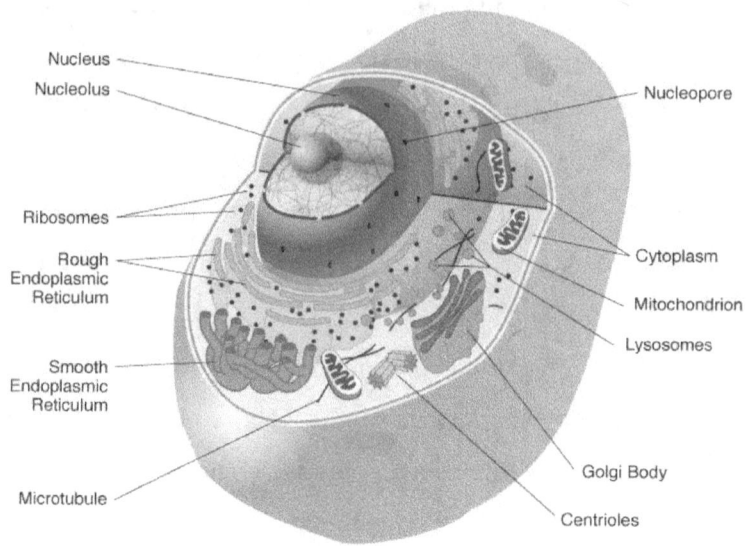

Nucleus
Nucleolus
Nucleopore
Ribosomes
Rough
Endoplasmic
Reticulum
Cytoplasm
Mitochondrion
Lysosomes
Smooth
Endoplasmic
Reticulum
Golgi Body
Microtubule
Centrioles

Figure 8.1 Diagram of the mitochondrion with labeled endoplasmic reticulum, and lysosomes [54]

If we magnify our single cell even more, we see that the organelles themselves are highly specialized and complicated. A single mitochondrion, no bigger than 10 micrometers in diameter, within a muscle cell, is wrapped in its own membrane, like the muscle cell itself; but, unlike our cell, it has two bilayers of membrane. Upon even closer examination, we notice something even stranger. The mitochondrion has its own DNA, and its DNA sequence is different from the DNA of the cell that it resides in. If we zoom out slightly to look at a large sample of the mitochondria within the cell, we observe that some of them are in the process of splitting and they are using their own circular DNA and proteins to do so. They are reproducing as if

160

they were bacteria. Strong evidence suggests that at one point, free-living bacteria joined together with or were engulfed by eukaryotic cells. The interaction fostered a cooperative symbiotic relationship. Now the two have become inseparable and mutually dependent. This type of cooperation, known as endosymbiosis, comes from the Greek *endo* meaning "within," *sym* meaning "together," and *bio* meaning "living." Modified bacteria are now living within our eukaryotic cells, and these bacteria-turned-organelles play a major role in the lives of our cells. The theory of endosymbiosis was first proposed by the Russian botanist Konstantin Mereschkowski, in 1905, but its most famous proponent was Lynn Margulis, who confirmed much of the theory through biochemical research in the 1960s.[55] [56]

According to endosymbiotic theory, an anaerobic ancestor of our eukaryotic cell engulfed a small, free-living aerobic bacterium. Instead of being digested completely, the bacterium not only managed to live inside the larger cell but thrived because of an abundance of food in the form of partially digested molecules floating in the cytoplasm of its new host. The bacterium took advantage of available oxygen molecules that were the toxic waste product of the anaerobic eukaryotic host cell, and metabolized them as molecular food, to form energy-rich ATP, a portion of the ATP seeped into the host cell. The host, in turn, used the ATP molecules to power its own cellular machinery. It was a win-win situation for both parties. In time, the engulfed bacterium became dependent upon its host for a number of cellular processes. The mitochondrion's DNA shrank as the host cell began to provide a larger portion of the proteins needed by its new guest. In short, the new mitochondrial guest provided energy via ATP, while the host cell supplied proteins, lipids, and protection from the external environment. There is also evidence that a very similar cooperative relationship emerged between a certain strain of cyanobacteria (blue-green algae), and the ancestors of eukaryotic plant cells. Endosymbiosis turned free living photosynthetic cyanobacteria into chloroplasts, the light-harnessing organelles of modern plants.

The evidence for endosymbiosis is overwhelming.[57] Mitochondria and chloroplasts both reproduce independently of the cell using their own DNA. The DNA of mitochondria itself is very close in sequence to the DNA of a group of free living bacteria called Alphaproteobacteria. Similarly, chloroplasts have DNA sequences that are very similar to those of a certain strain of cyanobacteria. The DNA of both organelles is circular like the original, engulfed bacteria, while the DNA within a eukaryotic cell is linear. The double-membrane of both chloroplasts and mitochondria are very similar in structure to the membranes of cyanobacteria and Alphaproteobacteria, both of which belong to a group of "Gram negative" bacteria with the same double membranes.[58] In addition, we have found organisms known as "living intermediates" that bridge the evolutionary gap between the organelles and their free-living ancestors. Moreover, ribosomes, which facilitate chemical reactions and communication within the cells, are also of bacterial origin.

8.2 Cooperation Within Our Cells: More for Less

One of the major differences between animal and plant cells, which are both eukaryotic, is that animal cells contain mitochondria and consume oxygen, whereas plant cells contain chloroplasts and produce oxygen. The evolutionary foundation of this process is that plant cells derived from cyanobacteria, which were oxygen producing, making the environment too toxic with high levels of oxygen and becoming the selective foundation of oxygen consuming eukaryotic cells. The anaerobic ancestor of our eukaryotic cells solved the problem by "domesticating" the free-living oxygen-consuming bacteria. The "domestication" increased the eukaryotic cell's complexity, and also increased its energy efficiency. An aerobic metabolism is 19 times more efficient than an anaerobic one. It also enabled the eukaryotic cells to survive in oxygen rich environments with access to more available Gibbs free energy, which would have been unavailable to anaerobic cells. Complexity, through the emergence of additional biological structures, can also be thought of as additions to the internal entrained biochemical complexity of pathways and information networks. Complex adaptive systems become more energy efficient and stable, and they provide a mechanism for the acquisition of available energy from the environment. The slower percolation of Gibbs free energy through the more complex system is the foundation of a much longer lifespan for eukaryotes compared with prokaryotic bacteria (24 hours vs. 30 minutes).

Of course, the precursors to mitochondria and chloroplasts were not engulfed intentionally, or with any foresight as to the cooperation that would emerge. The larger eukaryotic cell was simply following chemical signals that alerted it to the presence of food. Once the bacterium had been engulfed, new forms of thermodynamic exchange could begin, with new information embedded in each transfer of energy. This exchange caused the emergence of a more complex entrainment of new aerobic biochemical attractor pathways with the old anaerobic attractor pathways, leading to more energy efficient, robust, dynamic internal machines within the cell. It is an unintentional, physiochemical selection of the most energy efficient complex adaptive system. The integrity of this complex system is maintained through the capture of Gibbs free energy via the intake of food, which at its root came about through the capture of the sun's energy in plants and cyanobacteria and through plant food in other eukaryotic cells.

We also should remember that every one of our cells is really not a lone cell, but a team of interacting agents. In the same sense, a tissue is a "team" of the same type of cell, an organ is a team of tissues, and what we call a single body is in fact an enormously complicated swarm of individual, self-similar symbiotic units. In this chapter, we will explore questions related to the growth, maintenance, and self-recognition of our body: the swarm of communicating cooperators that constitute each of us. The evolutionary success of our cells is founded on their effective information and signaling capabilities, as well as on the stochastic compilation of complexity selected for its energy efficiency and robustness. Those systems that make it possible to allocate energetic resources in the most-efficient manner are selected to survive when energy is less abundant in the environment.

8.3 The Bottom-Up Embryo: A Process of Mini-Evolution

The emergence of various organs in multicellular organisms, such as humans, is the result of an array of coordinated activities among different specialized cell types: nerve cells, endocrine cells, and intestinal cells to name just a few. In each case, an emergent system has been selected for during the development of the embryo, based on the surface self-similarities of the cells constituting an organ. The final result is the emergence of complex organ systems, delicately communicative and cooperating with one another, sharing a self-similar chemical language, and thus giving rise to a new, emergent entity called the body: a macrostate embodying a long series of self-similar energy/information microstates.

The process of embryonic development mimics the entire macrostate of biological evolution within a very short

165

period of time. During embryogenesis, and later, organogenesis (the development of specialized organ systems), a growing animal receives Gibbs free energy from an outside source. The developing embryo is far from self-sustaining. Placental mammals receive their energy directly from the body of the mother. In birds, reptiles, amphibians, and fish, it comes from the battery-like power of egg yolk. Once an external energy source enables the reproduction of individual cells, the stage is set for a form of cellular evolution that takes place over the course of a single organism's embryonic development.

In typical Darwinian evolution, the diversity of living organisms is the fertile ground on which selection operates. Natural selection leads to the reproduction of surviving organisms, and the emergence of new species based on environmental factors. In the case of embryogenesis, the developing embryo faces a demanding nutritional environment. Here, differential success—the selection of certain cells over others—leads to the stepwise emergence of various organs, and ultimately, the emergence of a single, self-sufficient body. Billions of embryonic cells die and are thus selected against during the embryo's development—similar to the stalk cells of the slime mold and other "self-sacrificing" agents. Natural selection of the most energy efficient forms is at work, only at a much faster rate. Different cells in the body produce more offspring, depending on how much energy they use, and how much is available to metabolize. This mini evolution is

166

compressed in time to match the duration of embryonic development. But just as in the evolution of populations, here, replication, death, information transfer, and percolating Gibbs free energy result in the differentiation, specialization, and emergence of the complex adaptive system that is a newborn baby.

Involution, ingression, delamination, invagination, and polyinvagination are terms used to describe the "folding" of the embryo during development. The process of a developing body shaping itself is called morphogenesis. Certain zones of tissue sink into others as cells die, which rollup and seal the newly engulfed material. The developing embryo undergoes a terrifically coordinated shuffle of sliding cell types, inner and outer layers, flattenings, and engulfments, in order to select the most energetically efficient collection of cells as the final configuration. Watching a developing embryo is like witnessing the most intricate flower bloom.

Thus, embryogenesis is a process of the sequential expression of conserved genes throughout evolution. During embryogenesis, each step is briefly expressed, following the history of evolution to arrive at the final adaptive form. This is how, in embryonic development, many simple forms such as fish gills are sequentially observed, but are then absorbed, leading to increasingly complex forms. You may imagine a collection of compressed algorithms read sequentially leading to the manifestation of an efficient complex pattern. Simple programs are necessary, and form the foundation of stochastic selection, which in turn leads to more complex forms. Wolfram elegantly demonstrates the mathematical emergence of fern shapes from the simple interaction of numbers. We may imagine embryogenesis as a similar process of reading the embedded algorithms inside DNA—a process that reads from molecule to fish to man. The observation that ontogeny recapitulates phylogeny is like watching the translation of the compressed algorithm of DNA. So, in fact, all organs go through a compressed evolutionary expression in their development.

I speculate that this process of iteration is a mini evolution. The embryonic development and ontogeny of the immune system are examples of stochastic selection of the most energy efficient complex systems emerging in the microenvironment of embryo or maturing organisms. I may further suggest the fact that a large portion of the DNA, as much as 98 percent of the genome of many organisms, is not transcribed to proteins and is known as "junk DNA." However, this DNA has been mislabeled. In fact, it contains and transmits more useful information than has previously been recognized. These "junk DNA" codons, transposons, etc., may be in fact the historical library of evolutionary information that is conserved and reiterated and has useful functions throughout the growth and development of individuals. This DNA may also be a fertile bed for the stochastic emergence of new traits or even new species. These ideas, speculative as they may be, are consistent with the general theory of Thermoinfocomplexity, as it relates to the flow of information and energy within complex adaptive systems. In fact, Clune, et al., of Michigan State University have simulated the process of ontogeny recapitulates phylogeny in their evolutionary computer model.

8.4 Being Sticky and Self-Organized Complexity

In 1939, one of the pioneers of experimental embryology, Johannes Holtfreter, made an unusual discovery. He noticed that fragmented pieces of amphibian embryo, once dispersed in solution, would spontaneously adhere to one another. They would rearrange themselves into recognizable tissue types. In particular, the tissues that normally communicated through the self-similar language of their surface membrane structures came together spontaneously to form organs. They had been separated in space, but somehow, they formed exactly the assemblages you would expect to find in the developing embryo, even though they had been unmoored from their normal developmental processes. Certain tissue layers, like mesoderm and ectoderm, gather together to form the middle and outer layers of the developing embryo. Certain other types avoided one another, like the endoderm, or inner layer, and the ectoderm. Holtfreter named these preferences tissue affinities.[59]

The word "affinities" suggests that somehow, the various cells and tissue types were communicating and recognizing one another. Further experiments, undertaken later in the history of embryology, revealed that separated tissues showed two basic behaviors. Not only do cells aggregate with other cells of the same kind to form tissues, but the different tissues consistently end up in the same positions relative to one another. Individual cells are subjected to a shuffling process, and the emergent patterns are consistent. The mysterious (and effective) communication driving tissue affinity had to explain both of these features of the developing embryo.

Later experiments showed that one simple principle governed tissue affinities. It explained "like for like" and the spatial arrangement of the layers within the embryo. Its underlying cause did not require anything too fancy, either. There were no ultra-specific atoms, ions, or molecules that bound to a central atom with specific binding sites coding for certain tissue types; there were not even any overarching chemical gradients that governed the organization of the germ layers of the embryo, though gradients do form later during morphogenesis. The answer could be found in the simple physical property of adhesion, which derived directly from the physical makeup of the surfaces of the cells themselves.[60]

Each cell type is either more or less adhesive. The differential adhesiveness of cell types accounts for the spatial organization of the developing embryo. In other words, once you have a hierarchy of adhesiveness, you will notice that a few things happen automatically. Those cells with high adhesiveness will move toward one another over time. They will tend to clump together in the soup of the developing embryo. No chemical gradient brings them together; their surface proteins simply make them sticky, with one protein fitting into another and causing a chemical reaction that keeps them from pulling each other apart. The stickiest cells adhere to the stickiest cells to form a ball. Now imagine the "second-most sticky" cell type. Because of simple, physical laws, these cells will stick to the most adhesive cells with highest priority. This process maximizes the utilization of Gibbs free energy, just like a lipid bilayer does for the molecules within a cell. So, the second-most-sticky cells engulf the ball of the most-adhesive cells, now in the middle of the developing embryo. The remaining cells of the "second-most-sticky" cell type will stick to each other. We can expand this logic easily to include additional cell types of varied stickiness. A simple hierarchy results in complex, three-dimensional arrangements of tissue types. In this way, we witness the workings of a self-organizing system. Every cell type in the adhesiveness hierarchy follows one simple, thermodynamic rule: weaker adhesions are replaced by stronger ones, leading to higher levels of chemical stability. The structure always self-organizes, so that the total adhesive bonding energy between cells is maximized, resulting in the most stable structure.

Notice that an emergent, complex phenomenon—the budding embryo—can be explained by physical laws. The adhesion is the result of the interaction between self-similar surface polymers interacting with each other in an anti-parallel fashion, like a zipper. Differential adhesiveness results in specific tissue formation. Information transfer is at the root of morphogenesis. To emphasize the importance of this kind of signaling, researchers tried mixing non-biological fluids with otherwise biological embryo systems. The resulting spatial distribution was exactly what you would expect given the added substance's known surface tension. It did not matter whether the fluids were biological or not; what mattered was their surface tension. Far from there being an intrinsic embryonic force governing morphogenesis, let alone a top-down entity, embryo development is governed by a bottom-up, chemical signaling process. Cells communicate with local cells, and the result leads to the development of self-organized complex networks.

8.5 From Tadpole to Frog: Through Death, a New Life

There comes a time in every tadpole's life when it must become a frog. All those bodily changes—the development of legs and lungs, the loss of gills—require a greater amount of energy. At this time, the growing frog is still a lot like the developing embryo we first discussed: it is not yet sexually mature, and it must undergo major changes in its body's basic structure.

First, its body reabsorbs its tail. Shortly afterward, the legs begin to form. During the slow reabsorption process, the froglet, as it is known, eats less. Its tail acts as some sort of super sub-sandwich attached to the frog's body, which the froglet breaks down and metabolizes through the process of apoptosis.

Apoptosis, or "programmed cell death," is a process by which some cells within a body die, often to the benefit of the body as a whole—similar to how some cells in a slime mold die in the creation of the reproductive stalk. Apoptosis normally occurs when cells suffer injury due to mechanical damage, inflammation, or any number of other reasons. Cells undergoing apoptosis shrink, and develop what are known as *blebs*, or irregularities in the surface of the cell. The *blebs* cause the immune system to identify the cell as foreign, and to destroy the previously "self" cell.

Aside from injury, a cell may prepare for apoptosis by chemical cascade. Some substance within the body, but outside the cell, will attach to a surface protein sticking out from the surface of the cell. The surface protein changes shape, causing part of it to protrude into the interior of the cell. When the internal component of the protein changes shape, it triggers a reaction inside the cell, which in turn triggers another series of reactions. The resulting cascade of chemical information destroys the cell from the inside out. The cell shrinks, *blebs* form, and the immune system or other embryonic cells take care of the rest.

For the tadpole, this process is highly beneficial. When apoptosis results in cell death, the complex molecules that constitute it can now be broken down into simpler ones, becoming food for other cells. The cells die and their constituent parts are used as a meal for the rest of the body. Individual units within a body disintegrate or are eliminated as individuals. The result is increased protection, resources, and energy for the rest of the body. Moreover, the froglet saves a tremendous amount of energy, as it does not need to exert itself to find other sources of food during the transitional stage of its life.

This process of reabsorption is markedly similar to certain stages of embryogenesis. Because the tadpole is still developing, it could almost be considered an embryo with legs. As an embryo proceeds developmentally toward its adult body plan, certain parts of it die, while the rest of the body thrives. The body moves toward sexual maturity, and the fitness of the adult body increases. Reproduction, in this case, is possible only with the cooperation and even death of some cells. In the case of frogs, entire appendages are sacrificed. To the outside observer, these cells resemble armies of self-sacrificing altruistic heroes.

The same process of cell sacrifice occurs during morphogenesis in all animals. As we discussed, a hierarchy of adhesiveness is responsible for much of the self-organization of the developing embryo. As morphogenesis progresses, processes of apoptosis increase. The result forms a body. As cells die, limbs emerge. One eye turns into two. Fingers separate from one another, turning embryonic mittens into embryonic gloves. Cells die in rows, as toes peel away from each other. Growth and life are underway. Here, life and death are entangled through apoptosis and cell division. Swathes of individual cells die, and the body emerges as a cohesive, communicative, cooperating, symbiotic whole, complete with bacteria and mitochondria, all working together as one to make energy intake and utilization more efficient.

8.6 The Information Network of the Immune System: Local and Chemical

What sort of communication within a body is necessary to maintain its homeostasis? The virologist Frank Burnet asked, "How does the vertebrate organism recognize self from not-self in the immunological sense, and how did this capacity evolve?" [61] Along with Peter Medawar, Burnet won the Nobel Prize for Medicine in 1960, and these questions informed and guided his Nobel lecture. The answers have implications not just for the study of the immune system, but for the organization of complex life in general. Our immune cells and friendly bacteria defend us against usurping viruses, bacteria, toxins, and allergens, and they must distinguish between self and non-self in order to do so.

The immune system does not function in a top-down manner. Communication does not happen via one concentrated gateway, with a central node processing all lines of information. In the human body, as in most animal bodies, many dispersed chemical interactions add up to account for the process operating in the immune system. In other words, the information exchange of the immune response is entirely local and chemical in nature. When a cell turns tumorous, for example, or is infected by a virus, a change in its surface proteins alerts natural killer cells to target it for destruction. There is no central hub for all of our cells to pass through; they simply permeate among the cells of the body, either destroying them or ignoring them, depending on how the target cells are marked. Their interactions depend solely on the shapes and chemical interactions of their surface molecules.

The immune system began its evolution with selective permeability. In single-celled eukaryotes, the ancestors of our multicellular bodies, the primitive membrane acted as the gateway that received energy. Molecules with some sort of homology with the surface molecules of the cell membrane attached to the proteins, changing their conformation. Membrane proteins and phospholipids served as the gatekeepers. With the evolution of multicellularity, the division of labor and the specialization of cell types helped to create a more energy efficient, complex structure for the multicellular "colony" that composed the individual. The vertebrate immune system has evolved to become increasingly precise. The cells of our bodies tag unwanted foreign material for elimination, passing on information for how the agents of our immune system will treat them. Once basic rules are in place, the stage is set for the emergence of a body's overall immune response. The rules of self-recognition, spread across the individual agents, making it possible for new levels of organization to arise, without any top-down direction.

8.7 Hacking the Body and Crossing Wires

Drugs and poisons reveal the ways in which tiny alterations of the cellular communication system (often a minor molecular change) can have dramatic effects on the overall body. Emergent, complex systems are highly dependent on the webs of communication and information networks imbedded within them. For the slime molds, cAMP concentrations made a world of difference. As we will see later, ant colonies can be tricked into civil war with a simple molecular swap. A faulty identification system may cause the immune system to betray the body, for example in autoimmune diseases. In a similar way, drugs and poisons bond to cell receptors to take advantage of the body's preprogrammed responses.

Consider the powerful emergent effects of the parasite *Toxoplasma gondii*, which is present in about half the human population on Earth. Toxoplasma is a simple eukaryote whose lifecycle includes using rats and cats as carriers. Normally, rats are neurologically hardwired to avoid the smell of cat urine at all costs. In fact, researchers sometimes use cat urine to induce panic during studies of rat behavior. However, when Toxoplasma is present in a rat's brain, the behavior of the rat changes in an observable manner. Instead of scampering away from the fearsome smell of cat urine, rats infested with Toxoplasma show indifference or even affinity towards it, increasing the likelihood that the cat will eat them. Once consumed by the cat, Toxoplasma can continue its lifecycle and multiply in the body of its new feline host.

Other chemical mind-controllers exist in nature as well. The lancet fluke *Dicrocoelium dendriticum*, a parasitic flatworm, chemically impels its host, such as a hapless snail, to climb blades of grass. The exposed snail is much more likely to be snatched up by a hungry bird. The new, emergent behavior is devastating for the snail, but the fluke completes its lifecycle and spreads. A different fluke does the same thing to fish, causing them to splash at the surface of the water until they are eaten by wading birds. The hairworm forces grasshoppers to jump into bodies of water, where they then drown, allowing the parasite to swim away and reproduce. In every case, the behavior of the organism is radically changed by the chemical tampering of an internal parasite capable of manipulating the organism's information network.[62]

8.8 Broken Communication

Drugs and poisons are just one way in which changes in the body's communication system can have major effects on its functions. What about communication tweaks from within? Let us suppose an error occurs in our body, in which the immune system incorrectly identifies foreign material as being part of the body. With the body's defenses down, the invading material will proliferate unchallenged, and the unchecked spread of such harmful bacteria or viruses can result in disease or death.

But the acceptance of foreign material is not always a bad thing. Throughout an organism's lifespan, it accumulates what is called acquired tolerance to certain non-self material. An important example is the case of pregnancy. Although a growing fetus and placenta are not strictly "self," as far as the mother is concerned, they would not survive for long if the mother's immune system identified it as foreign. Over the course of its evolution, the immune system has acquired a tolerance to the growing fetus and placenta. Similarly, organ transplants cannot happen until the recipient body has developed an acquired tolerance for the incoming organ. Special drugs have been designed to induce tolerance and allow such operations to proceed.

The reverse situation, in which the body recognizes self as non-self, results in an autoimmune disease. Type-1 juvenile diabetes is one common autoimmune disease, in which the immune system destroys the beta cells in the body's pancreas, responsible for producing insulin. Without insulin, the body cannot utilize the sugar in its own blood. Glucose, the split product of common sugar, is the main agent of energy utilization within our body, especially for the brain. Without insulin injections on a regular basis, the inability to regulate blood sugar can quickly kill a diabetic.

Allergies are another way in which the body's signaling system can lead to an adverse emergent response. In an allergic reaction, the body overcompensates for something that it incorrectly identifies as dangerous. Because our immune system can respond rather drastically, its attempt to defend our body can sometimes kill, as in anaphylactic shock. This misperception occurs from a misreading of the information embedded in the molecules of the antigen, triggering an aggressive reaction.

The role of mucus in our nose is to protect membranes and organs from potentially harmful foreign intruders. When we breathe in an allergen, the mucus membranes catch the particles and trap them in their mucus secretions as a means by which to protect our tissue. Mucus is the collection of millions of antibodies secreted by the lining cells of our nose cells that self-sacrifice by ejecting their insides and combining their membranes into a viscous, allergen-trapping secretion. As airborne allergens collect in the nasal passage, they begin to irritate the cells in the deeper-tissue mucous membranes, until we sneeze, expelling the unwanted allergens and self-sacrificing immune cells along with them.

We are, in physical terms, a beautiful, bewildering, and humming network of information and energy transfer, competing and cooperating towards the most-efficient form that survives. By looking at the body as a complex swarm, from the perspective of an outside observer, we are able to see the intense amount of chemical interaction within our body's various systems. Information networks, as we observed with the immune system, represent the structural complexity of our bodies, and they help to maintain homeostasis. As demonstrated above, if something goes wrong with communication within the network, the effects can be devastating for the entire organism. In addition, information networks help to maintain our body's integrated structural complexity. The entrained and redundant paths in more-complex organisms provide alternative pathways within the information network.[63] For this reason, more-complex organisms are both robust and energy efficient. As we shall see in the following sections, more complex information networks allow the system to use less energy in maintaining its structure.

8.9 Metabolism: Santorio and His "Measuring Chair"

In 1636, Italian physician Santorio Santorii built an elaborate metal "measuring chair" hung from a balance beam. For thirty years, Santorio used it to measure his weight, the weight of everything he ate and drank, and even the weight of his urine and feces. He measured his weight after fasting, sleeping, sex, and work. He compared the meticulously derived values, and noticed that the weight of his urine and feces was always less than the weight of the food and water taken in. Santorio attributed the dissipation to what he called "insensible perspiration." Although today we do not think of the flow of energy and materials inside and through the body as "insensible perspiration," Santorio had conducted important empirical work.

181

Santorio had discovered the basis for the loss of heat and gases during the process we now call metabolism.[64]

Metabolism is the set of chemical reactions that sustain life. These reactions include glycolysis, oxidative phosphorylation, and the Krebs cycle. This generation of heat and gas, aside from perspiration, accounts for the bulk of the discrepancy between Santorio's food and water weights, and the weights of his excretions. As matter is broken down and reorganized within the body, it changes form in dramatic ways.

8.10 Basal Metabolic Rate: Kleiber's Scaling Law for Animals

The basal metabolic rate (BMR) is used to determine the metabolic efficiency of living systems. Measured in Watts, BMR is the amount of energy expended by an organism at rest. BMR measures the activity of a body's vital organs, including those of the nervous, digestive, cardiovascular, and endocrine systems. We use the value: BMR per gram of the body's tissue (BMR/g), in order to avoid the confusion brought on by direct comparisons of BMR among organisms of different sizes.

The first metabolic-rate scaling law was put forward in the 1930s, by Max Kleiber, a Swiss agricultural chemist who worked at UC Davis in the animal husbandry department. Kleiber derived his law by observing the metabolism of many different animals and graphing them in relation to their mass. His data showed a quantitative trend in the scaling of animal metabolism. The metabolic rate of an animal is proportional to its mass raised to the power of ¾. In other words, the higher the mass, the lower the BMR per gram. For example, the kangaroo rat, *Dipodomys deserti*, with a mass of 105.8 grams, has a BMR/g of 0.004887, while a bighorn sheep, *Ovis canadensis*, with a mass of 67 kg, has a BMR/g of 0.001711. The dromedary camel, *Camelus dromedarius*, with a mass of 407 kg has a BMR/g of 0.000552. Notice that although there is a near four thousand-fold increase in mass between the kangaroo rat and the camel, the metabolism of the kangaroo rat increases only by about nine times. We see the trend from smallest to largest, with BMR/g decreasing as body size increases.

However, when considering the world's largest mammal, we find the ratio to be different. The Asian elephant, *Elephas maximus* has a mass of 3.67 metric tons and a BMR/g of 0.000636, which does not seem to fit Kleiber's law.[65] Indeed, the Asian elephant has a higher BMR/g than the camel, meaning that although it is several times more massive, it is slightly less efficient in its use of energy. One possible reason for this discrepancy may result from specific environmental factors. The kangaroo rat, bighorn sheep, and camel are adapted to desert environments, where energy is scarce. With the scarcity of energy availability in the desert, there is a strong selective pressure for metabolic efficiency, translating into a low BMR/g. By contrast, the Asian elephant inhabits a lush, energy-dense forest, and as a result, the selective pressure toward an efficient metabolism is relatively weak by comparison. It would be interesting to investigate the complexity of the network of energy transfer within camels as compared to, for example, elephants.

8.11 Kleiber's Scaling Law on Bacteria and Protists

Despite a few outliers due to differential environmental pressures, as organisms grow bigger in size, in general they tend to become metabolically more efficient at a rate proportional to their mass raised to the power of ¾.

However, while the ¾ law is true for most animals and even plants, a different scaling law applies to protists and bacteria, organisms that are not a direct part of network-rich bodies. An exponent of 1 applies to protists, which means that protists scale less efficiently than multicellular animals. The metabolic efficiency of a protist is directly proportional to its mass, which means there is no efficiency benefit to a larger protist, as is the case with multicellular animals. An exponent greater than one is observed in bacteria, meaning bacteria—the simplest organisms—are the least efficient forms of life. As the mass of an individual bacterium increases, its metabolism increases dramatically, and the organism becomes progressively inefficient.[66] Perhaps for this reason, nature limits the size of bacteria (see *Figure 8.2a*).

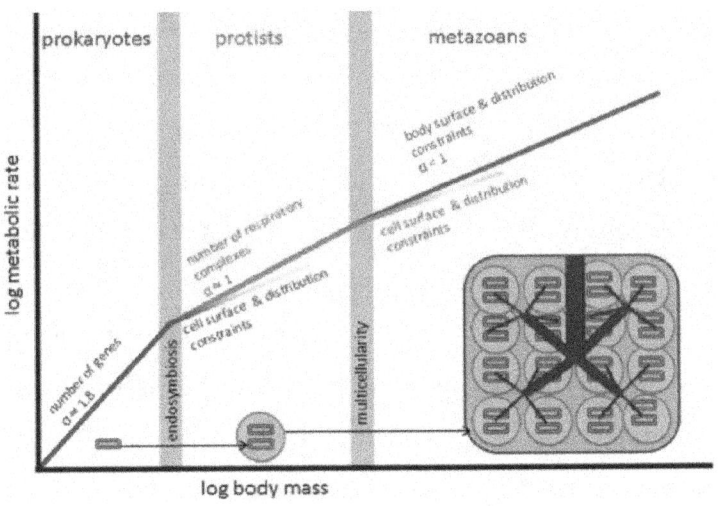

Figure 8.2a Relationship between whole organism metabolic rate and body mass for prokaryotes, protists, and metazoans plotted on logarithmic axes.[67]

In each case, the ratio of metabolic rate-to-mass depends on the specific structural constraints of the biology of each organism. The power laws arise as a mathematical description, characteristic of each of the emergent jumps in complexity in the prokaryotic cell, the unicellular eukaryote, and the multicellular organism. The exponent of the power law of BMR/g makes corresponding jumps (bacteria scale to 1.8, unicellular eukaryotes to 1, metazoans to ¾). In other words, it is the jump in life's complexity that leads to radical stepwise jumps in energetic efficiency, in terms of BMR/g. By looking at metabolic rates across all types of organisms, it becomes clear that energy expenditure (BMR/g) does not follow a single rule according to body size (e.g., humans). Rather, certain specific ratios arise from the specific organization of living things of varying complexity among prokaryotes, eukaryotes, or multicellular organisms. And this is why one can posit that the more-efficient extraction of Gibbs free energy into the environmentally selected, complex information-matter network of larger animals (size being a rough proxy for complexity), accounts for its metabolic efficiency. To put it simply, the more complex and intricate the network, the slower the dissipation of energy throughout. The organelles of the unicellular protist make the organism more efficient than the disorganized internal structure of the bacterium. But, beyond a certain point, greater jumps in efficiency require complex cooperative interactions with other cells. Energy efficiency, which does not necessarily correspond to an organism's mass, scales best with complexity.

Geoffrey West and others have argued that a branching circulatory system is at the root of the ¾ power law scaling in animals.[68] However, by focusing on the circulatory system alone, West neglects other entrained biological networks that contribute to metabolic efficiency, including biochemical networks, multicellular signaling networks of aggregated specialized cells, and the entrained networks of other specialized organ systems, such as the brain, liver, and kidney. Together the combined network effects contribute to a complex energy and information super-system. West is correct in that the microcosm of one organ system, the circulatory system, contributes to the fractal organization of the organism and is indeed often correlated with metabolic efficiency in animals. Nevertheless, the full explanation must take into account all the strange attractor pathways of the organism's information networks, including biochemical, cell-to-cell signaling, and the efficiency of communication arising within the panoply of interconnected metabolic networks. Energy efficiency results from the network effect, as evidenced by lower BMR/g. By treating the organism as a complex network of entrained biological systems, including, but not limited to, its circulatory system, we can explain scaling trends across all species: prokaryotes, protists, and multicellular eukaryotes included. Organisms lacking complex energy-saving networks are not as efficient as the more networked ones. Along these lines, we find the biological "economies of scale" at work. The branching tree of the circulatory system in a complex multi-cellular organism can be envisioned simply as a microstate involved in the distribution of Gibbs free energy (food), within the macrostate containing many other microstates for distribution of information and energy. For example, lymphatic system, liquid space between the cells, channels of information and energy exchange through the cell membranes, as well as, transfer of energy under the control of a multitude of communicating molecules, such as hormones, all within and between the specialized organ systems play a role. In other words, it is the complexity of the total network that dictates the metabolic efficiency. This fact explains why humans, which have much smaller mass than elephants, are

187

metabolically more efficient and have longer lifespans. We will discuss the relationship between metabolic efficiency and lifespans later. So, we can state that the macrostate system of information and energy distribution together contribute to the metabolic efficiency of complex organism, which we observe in ¾ power law.

8.12 Cyanobacteria: The Power of Specialized Parts

In order to fully understand increases in energy efficiency, we must take into account the various strategies of survival throughout particular ecological niches. By the stochastic processes mentioned in earlier chapters, organisms have evolved using a vast array of strategies.

One such outlier is the surprisingly efficient cyanobacteria. Cyanobacteria are a group of single-celled, photosynthetic bacteria that include the ancestor to modern plants and are thought to be the precursor to chloroplasts. Based on the general trends established by empirical evidence, we would expect the efficiency of cyanobacteria to be extremely low. However, in 2005 Makarieva, et al., surveyed the literature on bacterial metabolism, and found that certain cyanobacteria had metabolisms operating at rates that were remarkably similar to the metabolic rate of hibernating animals.[69] The key to their efficiency can be found in the variety of their specialized strategies of energy assimilation. The cyanobacteria *Microcoleus chthonoplastes* has four alternative chemical pathways for securing energy.[70] The first, aerobic respiration, is possible only in oxygen-rich environments, and it yields the most energy. The other three are various forms of fermentation, which occur anaerobically. One form takes place only when the molecule glycogen and the osomoprotectant glucosyl-glycerol are present. A second strategy uses elemental sulfur, and a third uses ferric iron, in addition to sulfur. In contrast, in the cells of more-complex organisms, there are only two pathways for cellular metabolism, and the cell will add the second only when oxygen is present. Energy options are surprisingly limited in these eukaryotic cells. The efficient cyanobacteria make up for their lack of an external communication network by having a high degree of internal complexity, as well as an internal signaling network with which to adapt to various environments. A multiplicity of chemical pathways can extract energy from a variety of environments, which enables the cyanobacteria to survive when resources of a particular type become scarce. These particular bacteria have the efficient power of complexity within.

The multicellular organism is complex in that its networks of cells relay information back and forth via hormones and chemical signals. Eukaryotic complexity is derived from interacting organelles with various chemical processes that allow for specific internal and external signaling. This outlier bacterial species may make up for its lack of external complexity by containing within its cell membrane a diversity of chemical metabolic pathways to react to various signals from the environment. It may lack the organelles of a eukaryote, but its complexity arises from an intricate network of choices within its metabolic pathways. The complexity of its entrained, varied, multiple biochemical strange attractor paths that attract self-similar molecules, results in robust and efficient energy use. The machinery shifts its gears so that energy may be harnessed with greater ease. In other words, the percolation of Gibbs free energy through it occurs more efficiently, and the dissipation of energy is slower.

To maintain their integrity and stability, complex adaptive systems require an input of energy from the outside, the availability of which is limited by environmental conditions. As a result, natural selection tends to favor more-energy efficient forms. We noted earlier that certain shapes, such as spheres or honeycombs, are more stable simply because of the energy efficiency that they confer upon the whole structure. In the same way, organisms with efficient metabolisms are selected for and will thrive over time better than their less efficient neighbors. Better communication networks with more-efficient information transmission through overlapping entrained cycles of complex metabolic networks, leads to more-efficient energy utilization. If we think of metabolism as the engine that drives and maintains every living cell, a cell with an efficient engine has a better chance of survival. It is better at interpreting signals that indicate how to capture useful molecules. It is better at processing those molecules into energy, building and replacing parts, and making new molecules to be used as fuel later on. On the other hand, a less efficient engine, one that cannot find messages in a sea of information, will be outcompeted by its neighbors. Just as certain stable geometric shapes are better at persisting in evolutionary time, organisms with more-efficient metabolisms in a changing environment will be better able to persist.

8.13 Mechanism of Power Laws in Allometric Scaling in Biology

Scientists have been measuring organisms' metabolic rate per gram as a way of comparing various species' metabolic efficiency. Geoffrey West and others have shown that larger animals tend to be more metabolically efficient, where the size of the animal correlates with its metabolic rate according to a ¾ exponent.[71] The accepted explanation is that this is due to the fractal structure of their branching circulatory systems. However, this explanation does not explain the various allometric scaling laws of bacteria, single-celled eukaryotes, sponges, and other organisms that lack circulatory systems.

These simpler organisms' metabolic rates scale according to different power laws. Single-celled eukaryotes have been observed to follow a scaling law with an exponent of one, while even simpler bacteria scale at an exponent greater than one. It is often stated that organisms get more efficient as their mass increases.[72] Although substantial amounts of empirical data support this assertion, very few mechanisms have been proposed to explain them.[73] Here, we propose a plausible explanation for metabolic scaling trends across all species, by using a bottom up approach.

At every level of complexity, life is organized into networks. Every living organism is an open system that exchanges energy with its environment and converts part of that energy into information captured in the chemical bonds of its information network. All cells use the Gibbs free energy they obtain from the environment to maintain homeostasis and reproduce. When chemical reactions come together in an organism, these reactions become dependent on one another and interact, i.e., the reactions become entrained. By chemical coupling together in a "chain" of reactions, the energy from one chemical reaction is caught and used in the next step of a metabolic process, harnessing the energy that may have otherwise dissipated into the environment as heat.

Over evolutionary time, network motifs have become more robust and interconnected. For example, in simpler organisms, such as bacteria, the proteins that the organism uses to interpret signals from the environment and exchange information are simple, but in eukaryotes they are much more complex. Likewise, prokaryotic transcription is simple. One enzyme binds to DNA, followed by several subunits that transcribe DNA into RNA by working down the line.[74] In eukaryotes, transcription is much more complicated. Introns must be removed, and the machinery involved in translation requires the movement of histones for proper gene regulation.[75] In prokaryotes, glycosylation is less complex and works by different mechanisms than in eukaryotes. In prokaryotes there are many different 'forms' of these proteins, because there has been no selective pressure for energy efficiency, and less entrainment with as many systems as you see in eukaryotes.[76]

Complex information networks carry adaptive benefits, allowing organisms to process information faster and more efficiently. This is known as the network effect. A random network, where interacting elements called "nodes" are connected to one another along "edges" at random, is seven times slower and far less robust than a real network (produced by natural selection) with the same number of nodes and edges.[77]

Whenever a new node is created, via mutation, that is inefficient, the organism is put at a selective disadvantage if energy (food) is less available in the environment. When a node adds to energy efficiency, however, it will be selected for during times of low energy in the environment (i.e., starvation). Complex networks allow for energy efficiency through robustness and fidelity. By becoming more intricate, cellular processes trap a larger portion of the information and energy brought into the cell. For example, bacteria ferment and eukaryotes respire. A fermenting bacterium may produce only two moles ATP per mole of glucose.[78] Yeast, a single-celled eukaryote, with its more complex, robust information network, made up of many entrained systems working together, can produce 32 moles ATP per mole of glucose. The response of the eukaryotic network allows the increased ATP production: glucose availability in the cell and a threshold concentration of ATP and ADP together trigger ATP synthesis.[79] Increased copying fidelity is another adaptive benefit of the complex eukaryotic information network. Different transcription enzymes have different rates of fidelity; the mutation rate in bacteria is ten times greater than in single celled eukaryotes, which is ten times greater than in invertebrates, which is ten times greater than in vertebrates. This fidelity allows for better information transfer, which decreases the amount of energy needed to fix mistakes.[80]

Large multicellular organisms tend to have the most complex information networks, and thus, have a lower metabolic rate per gram because they use, rather than dissipate, a larger fraction of the chemical energy they take in. For example, let's look at insulin production in humans. In multicellular organisms, the mechanisms that begin ATP synthesis are entrained with other chemical processes, allowing for storage of excess glucose before immediate use through insulin. With different specializations, a body is able to use glucose more efficiently through selective storage and use.[81]

A good metaphor for the passage of Gibbs free energy from the sun into the living organisms may be a pinball machine (*Figure 8.2b*). Energy (food) entering the mouth of the pinball machine. In bacteria it passes through a simple metabolic network. In a eukaryote, the path is more complex, and it takes more time for the energy to pass through the network as some of it is captured in the chemical structures in the amoeba. In humans the metabolic network is much, much more complex. It takes a long time for the energy captured in its structure to dissipate as heat. Bacteria has a reproduction lifetime of 30 minutes. Humans can survive without food for 30 days and they have a much longer reproductive time. This is because of humans extremely complex information and metabolic network. This is why bacteria metabolism follows a log power law with the exponent of 1.8, eukaryotes with an exponent of 1 and larger organisms including humans have a log exponent of ¾.

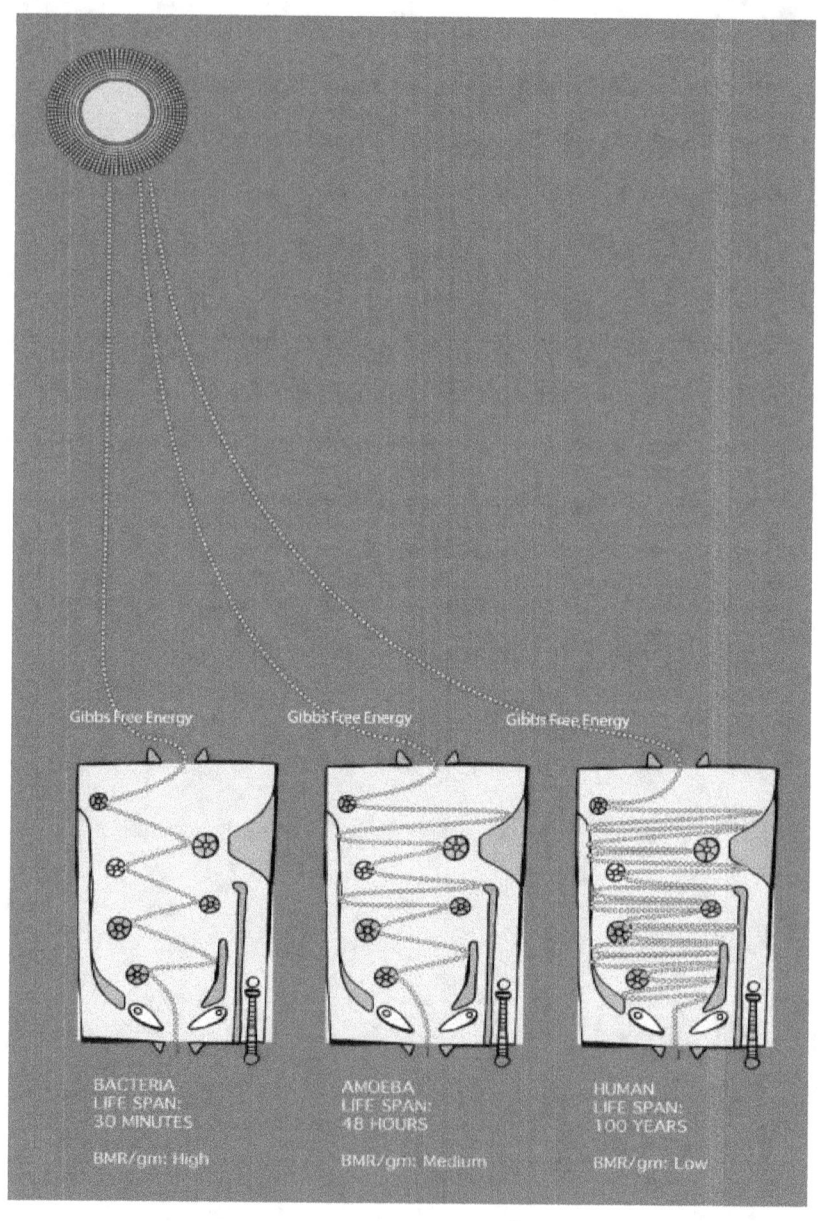

Figure 8.2b The metaphor of pinball machine for the passage of Gibbs free energy through simple and complex organisms.

The more that cellular mechanisms entrain in complex networks, the more efficient the organism becomes. Naturally, then, an organism with entire organ systems entrained together in a complex network will be more energy efficient than a simple, single celled eukaryote or bacterium. By keeping in mind what environment these organisms evolved in, we can explain why certain animals, like camels, have a lower metabolic rate than, for example, kangaroo rats, in the same environment. The answer to the power laws observed does not depend on body size; rather on the complexity of the organisms' entrained biochemical networks. By considering network effects produced by entrained chemical reactions, we gain insight into an evolutionary pattern of complexity and energy efficiency going hand in hand. Complex biological networks are more energy efficient. This fact explains the various allometric scaling laws. By connecting the principles of information theory, complexity theory, and thermodynamics in an evolutionary context, we achieve a plausible explanation for the observed phenomenon of various allometric power laws. Using these principles in the context of evolution could also help explain other interesting emergent phenomena across the progression of evolution, from molecule to man and superorganisms.[82]

Chapter 9 – Community in Transit

Figure 9.1 Birds flying in V-formation

9.1 The Logic Behind a Flock of Birds

Have you ever watched a large flock of birds swoop through the sky? The group moves as a whole through the air, as if it were operating with a single collective mind, or according to the directions of a single bird. We see the same phenomenon in schools of fish. Each individual fish moves in coordination with its neighbors, twisting and turning through the water in concert. One might be tempted to imagine this sort of cooperative movement is the result of careful practice or top-down leadership, yet birds, fish, insects, and even microorganisms group together without any executive direction.

Like all the emergent, self-organized systems we have discussed so far, flock behavior obeys a set of basic rules that each individual in a flock follows. The simplicity of these rules explains how bees, locusts, and other insects display organized group behavior. In 1986, computer graphics expert Craig Reynolds made a simulation of flocking behavior, known as *boids*. By making his boids fly according to a few simple rules, he was able to reproduce amazingly lifelike movement and coordination. While the computer simulation is not a perfect reproduction of nature, the algorithmic rules used by Reynolds helped to elucidate the logic behind how living organisms flock.

Birds flock according to the preference for energy efficiency. Flocking is adaptive for individuals as well as for the group and following this pattern has been selected for during the long history of bird migration. In flocking patterns, the first priority of the bird is to avoid midair collisions, which besides the potential harm, wastes precious energy needed for breeding and the remainder of the flight. Birds regulate the distance between each other to avoid this problem. This principle is known as *separation*, and it prevents accidents and overcrowding. A bird that follows the principle of separation minimizes its own energy expenditures and injuries. The behavior ensures the homeostasis of others in the group. Without this rule, the flocking pattern would crumble as birds

199

continually collided with each other. So, when programming a boid, one of the basic rules every boid should follow is separation.

Birds must also align their flight path with the bearings of their neighbors. This principle, known as *alignment*, allows birds to move collectively in the same general direction. In programming his boids, Reynolds added the principle of alignment alongside separation.

In order to avoid lagging behind and becoming a target for predators, flocking birds must maintain *cohesion*. Cohesion enables large flocks to keep their structure without dissolving. The push and pull rules of separation, alignment, and cohesion create a stable orientation for each bird relative to its flock. Reynolds programmed his boids with these three rules to replicate the emergence of flocking behavior. His simulation shows not just how birds flock in response to surrounding information, but how flocking emerges in general, as the most energy efficient and safe (physicochemically stable) formation possible.

A study published in *Science* described the neurochemical basis of flock formation. Researchers determined that mesotocin, a neurotransmitter, plays an active role in the creation of flock formation.[83] In different species of finches, a mesotocin-receptor distribution in the brain correlates with flock size. Mesotocin seems to influence what size flocks can be maintained. The administration of mesotocin increases social behavior, while a mesotocin antagonist reduces social behavior such as flock formation. The regulation of this chemical mechanism shows how complex flocking behavior can occur without central direction.

9.2 The Principle of Lowest-Possible Energy Use

While flocking, each organism reacts to local information supplied by its neighbors to determine how it should adjust its flight path, location in the flock, and direction. The result is a coherent, emergent structure with benefits in energy efficiency for the entire group. As a structure, the flock optimizes towards the lowest-possible energy configuration. The result is strikingly similar to how water molecules orient themselves toward one another in energetically efficient patterns. As water becomes a solid, the ice forms a rigid structure based on a low-energy conformation. The self-organization of birds in a flock is the result of a fundamentally similar process. Water molecules, like the birds in a flock are selected for based on the most-efficient way to use the input of energy necessary to maintain their particular structure. A study published in *Science* found that geese traveling in an organized, V-shaped pattern travel up to 70 percent farther than solo geese.[84] In another study, published in *Nature*, a flock of pelicans was trained to follow a motorboat and an ultra-light plane over a lake in a national park in Senegal. Each pelican was equipped with a bird-sized heart rate monitor, attached unobtrusively to their backs with adhesive tape. Pelicans traveling in the V-formation had heart rates that were 11.4 to 14.4 percent lower than pelicans flying solo.[85]

We see one of the most fascinating results of flocking behavior during the long flights of migratory birds. Each wingtip of a bird in flight creates a vortex. By flying in a V-formation, geese and other migratory birds capitalize on the vortex created by the bird in front of them. The vortex provides the birds with additional lift, allowing them to coast and flap their wings less frequently. By drafting the birds in front of them, the birds cut down on the air resistance over a long journey. Because the birds alternate flying at the front of the flock, a few bear the brunt of the wind, while the whole flock benefits collectively.

Figure 9.1 Birds flying in V-formation[86]

These same principles apply to professional cyclists in the Tour de France who form a tight group known as a peloton. This solid mass of bikers defines a constant speed throughout the entire race and winning each stage is often dependent upon breaking away from the main group, which requires a significantly greater amount of energy to achieve. In the same way, driving behind a semi-truck helps forge a path through the wind, improving one's gas mileage. The rule of energy conservation and drag reduction is not strictly biological; it is physical. When matter comes together, the stable, energy efficient arrangements are the ones that tend to persist, whether biological or not.

Like flocking birds, schooling fish conserve energy by reducing their friction in the water. The flock, schools, and swarms receive benefits that no individual organism could achieve alone. In the very short run, a bird might be struggling at the front of the flock's V, but over the course of a long journey, the bird uses significantly less energy. If every bird went its own way, it would have less energy for laying eggs and finding a mate and would need to eat more in order to sustain itself, stopping more often during an already long migration.

9.3 "The Wave" and the Network

Flocking birds have another mystery in store, one that seems to achieve the impossible. In 1984, Dr. Wayne Potts was studying a video of birds flocking, in an attempt to understand how the group moved with such conformity. He studied the dunlin, a type of shore bird with a black belly and beak, speckled brown dorsal feathers, and a proclivity for foraging and flocking along coastal mudflats and sandy beaches. The time it took each bird to respond to a visual stimulus in its local environment had been measured to be about 38 milliseconds. But, in studying videos of the dunlin, Potts found that the average time it took for each bird to respond to cues from its neighbor was only 15 milliseconds. Each bird seemed to be reacting faster than its own nervous system could relay information. The birds were responding to their neighbors' movements faster than their own physiological reaction time.

If you have ever been to a large sporting event, you are probably familiar with the stadium phenomenon of "the wave." Fans jump out of their seats in a huge, coordinated effort, giving the impression of a massive human wave flowing around the stadium. But, is each fan looking directly at his neighbor, waiting for a signal to know when he should join in? Of course not. If everyone did that, the wave would be relatively slow and unimpressive. Instead, each individual watches the wave coming from afar, timing his action with that motion to know when to jump up from the seat. The resulting pattern emerges from the simple actions of individuals practicing basic timing.

As Potts was able to show, birds are doing something similar when they flock together. Every bird in the flock initiates a "maneuver wave," which gains speed as it ripples through the flock. It starts slowly, but finally reaches average speeds three times higher than would be possible if the birds were simply reacting to their immediate neighbors.[87] Each individual bird watches the wave spreading throughout the flock and then times its own maneuver to fit perfectly into that wave. The result is the amazingly rapid, coordinated movement, no matter where the message originated from within the flock.

9.4 Schools and Swarms

Birds provide an excellent way of understanding flocking behavior, but they are by no means the only example of collectively efficient locomotion. Over 50 percent of fish species are known to swim in schools.[88] Schools of fish exhibit the same coordinated behavior as birds. And just like birds, they do it without a designated leader. Instead, each fish responds to a wide variety of movement cues from its neighbors. From the simple stimulus-and-response of each fish, the schooling phenomenon emerges.

Fish-schooling and bird-flocking evolved as totally separate phenomena. The sensory system in each is radically different; fish cannot focus their eyes directly forward because their eyes are located on the sides of their heads. They can, however, get an excellent picture of what is going on to the left and right, the areas of lateral movement on either side of them. This helps an individual fish see the movement of its neighbors, allowing it to move in concert with them. Some fish have what is called an "acoustico-lateralis" system, a biological apparatus composed of the inner ear and other sensory organs that runs laterally along the body. The lateral-line organs send information about water pressure and displacement to the brain, while the inner ear provides information about sound and gravity. This sensory system helps the fish to perceive messages, and to respond quickly to subtle changes in the water around it, furthering its ability to move with the collective.

Insect swarms contain a larger number of individuals than any group in nature. With billions of insects in the swarm, locusts in particular can cover hundreds of square miles. In 1952, I remember being in Shiraz, Iran, on a hot summer day.

Sitting indoors, I suddenly noticed that the sunlight had disappeared. I thought a storm cloud had passed overhead—an unusual event in the summer—and assumed that the repeated taps on the windows were heavy drops of rain. They were not. A swarm of locusts was passing through the city, plunging it into an artificial midnight. The locust swarm stretched for miles, and had traveled to Shiraz from Saudi Arabia.[89] Many locust swarms follow an activity pattern of boom and bust, commonly on a cycle of 13 to 17 years.[90] Because of the long lifecycle—alternating years of dormancy with one sudden appearance and breeding—predators cannot rely on them as a food source.[91] This ability to sate predators once every 13 to 17 years, and then breed in peace, is an example of one strange attractor's rhythmic cycle among many leading to an emergent, roughly regular swarm lifecycle.

Birds flocking, fish schooling, and insects swarming are examples of individual organisms forming and acting in groups. Each individual organism is a complex adaptive system on its own, but when a group is formed, that group becomes a new emergent system with its own properties and patterns, all selected to save energy. An emergent whole is not only greater than the sum of its parts, but different in kind. Groups form a new level of complexity with a unique structure, information network, and communication strategy. As we noted earlier, evolution is a stochastic process that selects for adaptive, complex systems that obey the laws of thermodynamics, communication, and complexity. Flocking, schooling, and swarming are stochastically selected through intricate information networking among individuals, resulting in the emergent structures that conserve energy per gram. However, emergent advantages in grouping are limited not just to energy conservation for an individual, but for the species itself.

9.5 Safety in Numbers

Perhaps the most influential factor in producing swarming behavior comes in the form of a threat. Both birds and fish coordinate their movements to help protect themselves from predators. Moving together in a large group gives each member a benefit it could not enjoy alone: the watchful eyes of the collective rather than the individual. If a predator moves toward a large group, its approach will be revealed at once. The danger inherent in losing group cohesion is most likely the evolutionary root for why maneuver waves begin in an inward direction rather than an outward one. As we noted before, the wave reaction time in a group is much faster than it would be in an individual's reaction time.

It is well-documented that smaller fish are much more likely to live out their lives in schools, while larger fish live more-solitary lives. This makes sense, given that smaller fish have more predators in the water, and increased danger acts as a selective pressure, encouraging schooling behavior. Any small fish that swims away from the school faces a much greater risk of becoming a meal. Research published in the *Environmental Biology of Fishes* shows that fish separated from their schools have higher rates of respiration than do those in schools, a fact caused by the increased energy expenditure of swimming alone (just as with the pelicans mentioned above).[92] The energy conservation of groups reminds us of the energy efficiency gained by multicellularity. The piranhas of the Amazon have also been shown to group together, mostly for defensive purposes.[93]

Swarming behavior improves the survival chances of insects as well. By observing Mormon crickets in 2005, researchers found a new way to determine the benefits of grouping together.[94] Mormon crickets are flightless insects that march in massive migratory bands across the western United States. By gluing tiny radio transmitters to the backs of these crickets, scientists were better able to study the fate of individual insects. They found that 50 to 60 percent of the insects that deviated away from the group by wandering off on their own were eaten over a period of two days. By comparison, none of the crickets that stuck with the group were eaten during the same two-day period. Predators had more difficulty finding a meal among the roving band of crickets than they did when they found an isolated individual. Here again, we see how information networks save lives.

Other organisms group together for protection from predators as well. Across the African savannah, zebra, gazelle, and wildebeest group together into large herds. Wildebeest even calve en masse, simultaneously giving birth to the entirety of the next generation within a few days. Without any design or intention, the herding network provides defensive benefits to its members by the emergent properties it creates. A large group increases the chance of spotting a predator early, and a predator's success depends on the difficult task of separating an individual from the group.[95] [96]

9.6 The Great Hunt

The same principles of collective defense also apply to collective predation. An organism hunting alone may do just fine. In fact, many species have evolved precisely this strategy, which has many benefits: an individual hunter does not share its prey with any companions, taking exclusive benefit in all of the nutrients its meal provides.

However, pack hunting often enables hunters to take down prey too big and strong to be killed alone. Consider wolves, which have evolved to hunt large animals, such as elk and moose. An individual wolf is unable to kill such large prey by itself because of the strength of the unforgiving hooves on its target. Even if one wolf could hunt and catch a moose, the wolf could be maimed or killed in the process. An injured wolf would go hungry very quickly.

The lone wolf rarely lives long enough to pass on its genes. Pack hunting is a different affair. Social hunting involves communication among the members of the pack, giving the wolves a better chance of killing large prey. Cooperative behavior allows wolves to expand their diet within their biological niche. Previously unattainable prey animals that provide more food become a viable option for a pack. Sure, the wolves have to share the food with one another but, in the end, there is more food to go around due to a greater diversity of prey. If there were always plenty of small animals for individual wolves to eat, perhaps pack hunting might not have evolved, but when the supply of small animals dwindles, pack hunting provides a clear selective advantage. Here we see the emergence of cooperative behavior and herding when Gibbs free energy in the form of prey is scarce. When there is plenty of small animals for food, we see individuals disperse. We observe the same tendencies toward cooperative group action and individualism with quorum-seeking bacteria, slime molds, and later, with human societies.

Dolphins also form groups similar to the wolf pack. Feeding mostly on small, fast fish, dolphins have evolved social hunting behavior to conserve energy. A single dolphin chasing a fish will have limited success; its prey can move quickly in the water and make sharper turns. To overcome the schooling of large numbers of small fish, dolphins must also act cooperatively. A group of dolphins herds a school of fish into a particular place in the water, and then circles the fish to keep them densely packed together. Individuals then take turns swimming through the school and eating the encircled fish. In this way, each individual benefits from the collective work of the group. Moreover, this technique demonstrates how one species' adaptive strategy towards energy efficiency causes its predators to adapt a similar cooperative behavior.

We find similar pack-hunting techniques in response to herding prey throughout the animal kingdom. Lions, hyenas, and African wild dogs have adopted this strategy in response to the collective groupings of their prey. We humans have also evolved pack-hunting behavior, using our strength in numbers to surround prey that would otherwise have outrun or overpowered us.

Pack hunting also increases the likelihood of finding potential prey. A large school of predatory fish is much harder for potential prey to avoid. Instead of a single predatory fish, the prey must now escape an entire mob. Atlantic herring hunt in packs, and they feed on a tiny species of crustacean, known as copepods. When approached by a fish, a copepod will dexterously "jump" out of the way, detecting approaching pressure waves in the water with its extended antennae. Groups of herrings have evolved a strategy for dealing with their quick prey: together, they swim toward a school of copepods in a formation called "ram feeding." The herring approach the copepods with their mouths open, spaced apart from one another at exactly the same distance that an individual copepod will jump when threatened. In this way, a copepod jumping to avoid one herring will land directly in the open mouth of another herring. An individual herring could never catch a copepod, as copepods can jump up to 80 times in a row before tiring.[97] But together, they act as a net. The school of herrings cooperates to take down prey, greatly enhancing the group's energy efficiency and effectiveness as hunters.

In another example, a group of researchers analyzed aerial photographs of foraging groups of Atlantic bluefin tuna and found that the school was shaped like a parabola that fanned out in front. This is the optimal configuration for foraging and hunting, bringing in prey like a funnel, as though the school itself possessed a single, gigantic mouth.[98] Foraging fish also receive the energy-saving benefits from the network effect, by sharing information about the surroundings among themselves. When one tuna sees its neighbor feeding, it automatically begins a foraging pattern. In this way, fish share the information that food is present, and the group can take advantage of a signal that propagates quickly throughout the school.[99]

9.7 Local Rules, New Complexity

Flying, swimming, or crawling together has benefits, including protection from predators, collective predation, and reproductive advantages. Together, individuals spend less energy in the effort to secure a meal, evade a predator, or to simply get from point A to point B. Group dynamics comes with emergent layers of complexity and enhanced communication.

Over time, the more-energy-efficient forms will have specific advantages over less-efficient forms, and they are therefore selected to persist.

At the micro level, communication usually happens via chemicals. On the macro level, we see that communication takes other forms: sound, sight, touch, smell, or even stranger modes such as the lateral line sensors of fish. Think of all the ways in which we exchange information with other human beings. Our hearing allows us to detect not only language, but the sound of water, the cries of predators and prey, and an approaching avalanche. Our eyes are capable of seeing a wide variety of threats and opportunities in our surrounding environment. We can identify potential meals and tease out the subtleties in the facial expressions of our fellow humans. We can smell food, smoke, other humans, decomposing matter, and all manner of other useful inputs. Our nervous system allows us to experience pleasure and pain. Compared to a slime mold, we are a Swiss army knife of information interpretation. Different senses allow for different and varied types of communication tools and responses. To be considered information, all these inputs must be converted by the organism receiving them, into electrochemical messages. Communication at the macro level is still dependent on communication at the micro level. In the end, the interplay of energy and information among self-similar agents from molecules to bacteria, birds, fish, wolves, and humans results in energy efficient complex patterns of behavior.

214

These complex patterns cycle within strange attractor patterns. Strange attractors, for our purposes here, may be defined as a path derived by a number of not clearly defined forces that collectively act upon the agents constituting a complex system, for example a swarm of locusts. Following this path results in a cyclic or rhythmic pattern of emergence.

As we have seen, macrostate communication contains within it numerous fractal microstates from sender to receiver, creating a single pattern of behavior. Birds flock, fish school, wolves hunt. The emergences required for such phenomena are tiered: the top levels require the lower levels in order to function. It is a buildup of stochastic processes. In this way, complex systems build upon one another in nature. The story of flux among various emergent stable systems is the story of evolution on Earth. Some arise, some peter out, some are selected for and some persist. But the more things change, the more they stay the same. These new forms of information transfer still provide the basis for self-organizing systems to emerge from the relative simplicity of organisms acting on their own, all according to a handful of physical rules that lead to the emergence of new complexities, as driven by energy scarcity.

Chapter 10 –
Superorganisms

Figure 10.1 Honeypot ants

I was eight years old on summer vacation with nothing to do, when my brother and I noticed two different types of ants in our backyard. Each kind emerged from separate holes in the ground. One type was red with big pincers, and much larger than the other. The others were black, with smaller, but sharper pincers. Being young boys with little to do, we wanted to see if we could make the ants fight. So, we held one red ant and one black ant pincer to pincer, and when they became aggressive, we placed them next to each other on the ground. The ants wrestled for a time, but then stopped abruptly, returning to their nests on opposite corners of the yard.

Suddenly, swarms of ants came out from their respective nests. They climbed over each other in a frenzy. Both the red and the black ants were ganging up on individual ants and taking them back to their nests, where they were probably chopped apart. The war continued for several days. Mesmerized, my brother and I watched for hours and hours.

We then decided to see what would happen if we put breadcrumbs out for the ants—if they would come for the food and stop fighting. One black ant ran to the bread and then back to the nest. Then a whole bunch of black ants came and took the bigger pieces of bread and rushed back inside the nest. I noticed something on the cement where the ant had just walked; a glistening line ran across it. I saw that the ants were actually secreting something from their tail gland. So, we started playing games with these ants. When they went back to the nest, we wiped the line away. They marched back from the nest toward the bread but accumulated at the point where the line had been wiped, wandering around in confusion. The war stopped.

So, then my brother and I simply restarted the war once again by touching the red and black pincers to each other. They started wrestling. After the wrestling, they went back to their nests and the armies returned and resumed fighting. We became quite interested in seeing which ants would win the war. We did this all summer. I had claimed the army of the black ants, which were greater in numbers, and my brother claimed the red ants, which were individually larger and stronger. Then something very puzzling happened—the red ants started to dig and make a tunnel underground, which opened up a slippery sand funnel about three centimeters from the entrance to the black ants' nest. Whenever a black ant would start to leave its nest, it would be caught by a watching red ant, thrown into the trap, and quickly taken away by a group of red ants. Where, we did not know. The red ants had strategized to make this underground tunnel and funnel system, and since we could not see beneath the surface, whenever a black ant fell, it simply disappeared. Soon, another black ant would emerge from its nest and again would fall into the opening. The same thing happened again and again.

Then something very interesting occurred. We noticed that the black ants had closed the old entrance to their nest and relocated it far away, near the wall. They could now leave the nest without encountering the enemy. My black ants were greater in number and capable of adapting their strategy in real time. I was hopeful that they would emerge as the clear victors. Then, to my chagrin, I came outside the next morning and saw that my younger brother had taken a garden hose, put it into the entrance to the black ants' nest, and flooded it. All the black ants had been killed. The war was over.

10.1 Armies Without a General, Queen Without a Monarchy

Each ant selflessly marches to war against its enemy without much consideration for its own individual well-being. We might swat whole legions of ants dead, but they keep coming, undeterred by our size and might. The individuals die, but the colony lives on. At first glance, these sacrifices appear to be heroic altruism, in the popular understanding of the term. The individuals seem to act as martyrs for the greater community. But what appears to be altruism often has little to do with the organism's conscious intentions. For ants, biological altruism applies not just to wartime, but to the resource economy of the colony. In fact, much of the time and energy of the ants in the colony is devoted to only one particular ant dubbed by entomologists as "the queen." However, an ant colony is no monarchy. The queen does not direct orders to her subordinates. No single ant coordinates the complex social structure of the colony. The incredible amount of order and unintended altruism occurs from the ground up, just as in slime molds or flocks of birds. From the countless variations in signaling and response of individual "agents," one coordinated, meaningful action results.

Charles Darwin remarked on the incredible specialization of cooperating ants in *On the Origin of Species*. He described the sterile workers and soldiers of *Cryptocerus*, as well as the drones of *Myrmecocystus* that never leave their nest.[100] Since Darwin, the study of ants has seen a renaissance. We now know quite a lot about the ant colony's intricate division of labor: the minuscule "minims" that tend to the queen and her developing young, the drones, and the huge workers known as "majors" that excavate and defend. It seems as though there is a type of ant for every job, the different types interacting with each other in the same colony. As specialization increases, so does the complexity of the hive system and its network of communication, resulting in greater energy efficiency.

10.2 Specialization at Work: Fungus Gardens, Slavemakers, and Honeypots

In some cases, ants specialize in interacting with other species. Certain varieties domesticate aphids and use them like humans use dairy cattle. Both groups benefit. The ants protect the aphid eggs, and in turn, the aphids supply the ants with a sweet food called honeydew. Each species sees greater reproductive success because of this mutual cooperation. Ant species have also developed mutually beneficial relationships with fungi. "Leafcutter" ants are known to actually cultivate entire fungus gardens, tending the fungus over time and defending it from predators. The fungus could not survive without the ants, and the ants, in turn, rely on their crop of delicious fungus for food. When a young queen leafcutter ant is ready to start her own colony, she carries a piece of the fungus with her and lets it grow in the new nest. In nature, the reach of cooperative behavior often extends across the species boundary.

Other ant species take a less harmonious stance toward outsiders. In *On the Origin of Species*, Darwin described a species of slavemaker ants whose members do very little work themselves. The warrior caste brings in labor from the outside by stealing pupae from other colonies. When the pupae mature, they become the slaves of their captors, working as though they were members of the colony. The slavemaker ants show very little interest toward any task other than raiding. Darwin wrote that "they are incapable of making their own nests, or of feeding their own larvae. When the old nest is found inconvenient and they have to migrate, the slaves determine the migration and actually carry their masters in their jaws."[101] This species of slavemaker ants are "obligatory slavemakers," which means they are dependent on the labor of captive ants to survive.[102]

Jean Pierre Huber, a scientist whose work Darwin referred to in *On the Origin of Species*, performed an experiment in which he provided a group of slavemaker ants with food, water, and all the resources they needed to live, but deprived them of their slaves. The slavemaker ants quickly began dying off. Darwin wrote, "[s]o utterly helpless are the masters, that when Huber shut up thirty of them without a slave, but plenty of food which they liked best, and with their own larvae and pupae to stimulate them to work, they did nothing; they could not even feed themselves, and many perished of hunger." Eventually, a

221

single slave was introduced "and she instantly set to work, fed and saved the survivors; made some cells and tended the larvae, and put all to rights."[103] Although not all slavemaker ants are so dependent, this relationship is an example of absolute dependence and specialization. Yet, imagine how much energy the slavemaker ants save by using other ants.

There is one ant even more specialized than any mentioned thus far. As we have seen, different varieties of ants range the gamut in their interactions with other species, from mutually beneficial, symbiotic relationships to strict exploitation. There is perhaps no stranger ant, however, than the "honeypot." In terms of its specialization, it has to be one of the greatest team players in all of nature. A honeypot ant is just what it sounds like—an ant that acts as a storage container of honey for the rest of the hive. They hang in clusters underground, in the colony. They do not move; these storage ants are fed by worker ants until they swell up to be globes about the size of grapes.[104]

Figure 10.1 Honeypot ants[105]

Then they wait and feed the other ants. Soon these honeypots are so stuffed that they can no longer fit through the tunnels to leave the hive. As with all ants other than the queen, these ants completely eschew their own personal chances of reproduction in the service of the group.

10.3 Cooperation Beyond Self-Interest: A Paradox?

Although nature is organized largely by cooperation through communication, the traditional view of biological evolution predicts that most individuals will act according to their own individual self-interest. They compete to pass on their personal genetic information. As we have seen, however, ants tell a different story. Aside from the honeypot, other castes of ants in the same colony forage and tend larvae. The honeypots act as the hive's communal food supply and nothing more. The reproductive success of a honeypot ant is nonexistent. How could such an extreme specialist that is not much more than a storage container have evolved? The problem seems to be especially vexing when we consider that, as sterile castes, all specialized ants can never pass on their genes. If evolution is based on the inheritance of traits, how can evolution take place once inheritance is out of the equation? The answer is that the entire ant colony, with its queen capable of passing on the genes of all ants, functions together as one superorganism.

10.4 Kin Selection: Selfishness and Altruism are Interrelated

Bees, ants, wasps, and termites have been at the center of arguments surrounding Darwinian evolution. Some of the most novel and intriguing theories of selection have come from the field of entomology, as have many of the most innovative thinkers in evolutionary theory. E.O. Wilson's controversial 1975 book, *Sociobiology: The New Synthesis*, still a focus of heated debate in biological and social circles, has its theoretical foundation in his many years studying ants.[106] So many of evolution's most brilliant theorists have focused on insects—the troublesome organisms whose sterile castes Darwin once considered "fatal to the whole theory." We ask, how could sterile ants have evolved if the genes of the sterile individuals are never passed down?

224

Darwin's own difficulty regarding how these sterile insects could have evolved was related to a missing piece in his view of the biological world, one that led him to place too much emphasis upon the individual as the unit of selection. He never understood the mechanism for inheritance. Unbeknownst to Darwin, around the same time, Gregor Mendel was formulating his theory of genetics. Darwin's theories gained widespread notoriety almost immediately, while Mendel's work was largely ignored during his lifetime, lost to obscurity until it was rediscovered in the 20th century. Darwin never even heard the word "gene," and so the missing piece eluded him.

Without any knowledge of how traits pass from one generation to the next, Darwin struggled to reconcile predictions of his theory with a few apparent contradictions in the real world. One such mystery was the case of sterile insects. How could neuter insects have evolved if natural selection depended on inheritance, and neuter insects never had offspring of their own? The dawn of genetics in the last century, however, provided biologists with the mechanism for explaining how Darwin's theory actually worked.

Richard Dawkins is one writer whose work attempts to explain the paradox of Darwin's neuter insects. In his 1976 book, *The Selfish Gene*, he posed a question: "Evolution works by natural selection, and natural selection means the differential survival of the 'fittest.' But are we talking about the fittest individuals, the fittest races, the fittest species, or what? For

225

some purposes this does not greatly matter, but when we are talking about communication, cooperation, and altruism it is obviously crucial."[107] What unit does natural selection truly act upon? Dawkins' answer is that the only part of an organism that actually makes it from one generation to the next is its information-rich genes, the molecular code in each cell upon which all life is built. In this light, natural selection actually means the survival of the fittest genes, as they have interacted with the environment to give rise to better or worse adapted phenotypes, rather than the survival of the fittest individual organisms. Genes, the fundamental replicating unit of all life, along with the bodies they reside in, are modified by the evolutionary processes over time. According to Dawkins, selection acts only at the level of the gene.

I would like to dispel confusion in the public mind that Dawkins has created by confusing the self-interest of an individual with the selfish gene. The gene itself is not selfish. Rather, the individual harboring the gene serves its own self-interest as a living complex adaptive system. First, I feel compelled to state that a single individual's genes in all living organisms that reproduce sexually (99% of living organisms) is not the unit of selection. The unit of gene selection is a combined haploid gene from a male and female. Therefore, if we are to attribute selfishness to the individuals harboring the selectable genes, we should call it a "selfish complex." However, these types of metaphors do not elucidate the process

226

of adaptive gene selection. It is extremely important to understand the evolutionary advantages of male and female gametes coming together to transfer their selected genes based on variability as the ground for selection of adaptive progeny.

Though all the worker ants remain sterile, by helping the queen to lay more eggs, gathering her food, cleaning the nest, and defending the colony from predators, they are actually improving their own genes' chances of making it to the next generation. Of course, ants are unaware that their genetically predisposed behavior helps the reproducing queen, which thus passes their own genes on to the next generation.

When an individual helps its relatives, it also helps copies of its own genes residing in its relatives' bodies. This phenomenon has come to be known as kin selection, as first put forward by R.A. Fisher and J.B.S. Haldane, but developed mathematically by W.D. Hamilton.[108] Kin selection describes the evolution of behaviors that benefit the genetic relatives of individuals, even at a personal cost. When should we expect such behaviors to have evolved? Hamilton offered a mathematical answer to that question, which became known as Hamilton's rule. It states that, given three variables, r (genetic relatedness between the interacting agents, where one appears to act altruistically toward the other), B (the reproductive benefit to the recipient of the altruistic act), and C (the reproductive cost to the altruistic agent), r multiplied by B must be greater than C (r x B > C). This formula predicts that the closer the genetic

227

relationship between two agents, the more we ought to expect altruism to have evolved as an evolutionarily stable strategy. It implies that selfishness and cooperation can be two sides of the same coin, depending on what agent the human observer deems is the helper and what agent is the helped. If we look at a mother bear defending her cub in terms of individuals, we say that the mother is acting altruistically. If we look at it in terms of genes, we say that the mother is acting selfishly, that she is ensuring her genes survive into the next generation. In this view, altruism and selfishness are interrelated.

10.5 Beyond Kin: Gene Transfer Unit and Group Selection

Hamilton's explanation for *eusociality* (defined as living in a cooperative group with usually one fertile female, several reproductively active males, and many more sterile females) in insects relied on the abnormally high genetic relatedness of sisters within the colony. This high relatedness of ants in a colony drives them to place a higher priority on aiding the queen in reproduction than on producing their own offspring. However, eusociality has also been discovered in a variety of other species without the extreme genetic similarity of the ant colony. These species included naked mole rats, termites, aphids, and snapping shrimp.[109] Cooperation pervades nature.

We do not need to look to genes as the mechanism of causation. It is true that the genes of ant colonies get passed on to the next generation, but so does a copy of the colony itself.

228

We can look at the ant colony as a superorganism, where individual specialized ants act like individual, specialized cells.

The arrangement is very similar to our own body. Our somatic cells—liver, blood, kidney, brain—are all sterile, but their genes are passed on by reproduction through our gametes. In that sense, when a man and a woman come together to reproduce, they can be thought of as a *gene-transfer unit*: a biological system that reproduces and passes on its genes to the progeny of that system. The genes of specialized honeypot, soldier, and farmer ants are all passed through the queen, so we may actually think of the whole colony as an "individual" reproducing unit, similar to the way we might think of ourselves plus our mates. The large separation in space between ants does not make a single, specialized ant any more truly "individual" than does a blood cell or liver cell in our body. In either case, the constituent units communicate and collaborate, ensuring the survival of the respective gene-transfer unit. This unit, in humans, is the superorganism of two reproducing individuals, and in ants it is the fertilized gamete. As individual humans have haploid germ cells that contain only half a complete human genome, a human gene transfer unit must be a man plus a woman to create a child. Just as our fat cells store fat as a source of food for other cells, honeypot ants serve the same function for the colony. They are stores of food for other hard-working cells, which act as the worker ants.

The ant colony is a superorganism and its combined unique package of phenotypic characteristics is subject to evolutionary selection. The variants arise from the combined gene transfer units and are selected to produce a new form, whether in an ant colony or in a pair of sexually reproducing multicellular organisms. For ants, the group, acting as one individual, is the unit of selection. Here, evolution selects the most energy-efficient form, whether an ant colony or two individual humans, both of which house and transfer their genes to the next generation. Though Darwin was never aware of the existence of genes, he was able to correctly identify group selection as the means of gene transfer into the next generation. He writes, in *On the Origin of Species*:

> "How the workers have been rendered sterile is a difficulty; but not much greater than that of any other striking modification of structure; for it can be shown that some insects and other articulate animals in a state of nature occasionally become sterile and if such insects had been social, and it had been profitable to the community that a number should have been annually born capable of work, but incapable of procreation, I can see no especial difficulty in this having been effected through natural selection."[110]

Darwin understood the importance of group selection: the idea that nature selects the fittest groups, not just the fittest individuals. Because he did not know about Mendelian genetics, he did not understand how sterile castes inherited these traits. Regardless, Darwin understood that not just individuals, but groups, competed and cooperated for resources in their ever-changing environments. Groups could be responsible for carrying the torch into the next generation. Of course, no ant can identify its relatives. It is also unaware that its own genes are housed in the queen. "Altruism" and "cooperation," as we have stated, are concepts invented by the human observer. The root of these behaviors lies in energy conservation through information networking, selection, and the emergence of energy-efficient complexity in organisms, superorganisms, or any other complex adaptive system.

10.6 Pheromones

A colony of ants appears to move fluidly as one entity, just as humans walk as a colony of cells. E. O. Wilson chose the word "superorganism" to describe a group of tightly cooperating, yet spatially distinct animals. In the superorganism, each individual acts in its caste-specific role, seemingly for the good of the colony. The workers cannot pass on their genes without the queen, and the queen cannot eat or take care of herself without the workers; the colony would be defenseless without the warrior caste, the eggs will not hatch without the minims, and the honeypots cannot do anything on their own. The colony is so complex and interdependent that it is as if it were one organism formed from the bodies of many. Different, specific castes are like different organs of the same body. The genetic material of the superorganism just happens to be spread among separate "cells" like the bodies of individual ants. Communication and information transfer between the separate individuals may occur across longer distances compared with the communication between cells in a multicellular organism but, in any case, information is being transferred. This information is necessary for maintaining the structural stability, or homeostasis, of the organism or colony, as well as ensuring the survival of its gene transfer unit. It is the extensive information networking and the complexity of the colony that has led to its energy efficiency as a basis for its evolutionary selection. This is why ants have emerged and persisted for over 110 million years and constitute 15 to 25 percent of the Earth's biomass.[111] [112] Here we may visualize an individual ant as a macrostate of attractor basins housing the collective energy of numerous microstates of fractal cells. On another level, we may visualize the colony as a macrostate housing the collection of energy/information contained in the microstates of constituent individual ants. Each macrostate results from the selection of the most-energy-efficient complex forms during periods of energy scarcity.

We note that there is no major difference between a colony of ants and the human body. When we go for a walk, we are practicing a complicated dance involving multiple body systems. Nerves fire, muscles contract, signals from our brains keep our limbs moving. We digest our food and our heart continues to beat, pumping blood that gives oxygen to our muscle cells. We make tiny adjustments in our gait when we see something on the ground we would rather not step on, such as an ant. The point being, most of the body's communication is coordinated through the central nervous system, as well as through chemical signaling between individual cells. The ant colony superorganism has extensive chemical signaling, but no central nervous system to coordinate actions with the same quickness. It is one thing to describe a superorganism as an organism whose genes just happen to be spread across several bodies, but it is something entirely different to determine how, exactly, those disparate bodies can be said to act together as a whole. The nervous network of ants is a distributed one, with each ant boasting an average of 250,000 neurons directing its own actions. As we have said before, a necessary condition for the emergence of a new level of complexity is efficient and effective information transfer between the individual parts, whether molecules or ants.

How do ants transfer information among each other? Most of their communication is chemical in nature, and most of these chemicals contain very simple bits of information. Certain chemicals tell ants to follow, another identifies ants as part of a particular colony. Pheromones serve as the primary means of communication between ants and within ant colonies. As pheromones would escape randomly into the air, ants need direct contact to pick up the correct signal. An ant's antennae contain olfactory receptors comparable to those in our nose. When the antennae encounter a pheromone, the chemical binds to the olfactory receptors and relays information to the ant's tiny brain, which directs its behavior. The ants are communicating, but the simplicity of the ant nervous system (ranging from 10,000 to 250,000 neurons, as compared with a hundred billion in the human brain) means that its communications and resultant behaviors are also relatively simple.[113] The same is true for slime molds, though they respond to a different chemical signal: cAMP. An ant touches a chemical, the chemical binds to the receptor, and a chemical cascade is initiated within the ant's body. This process is only one step removed from the chemical communication of the slime mold, in that a cascade must occur after the ant makes contact with the pheromone. As we have described in Chapter One, the information communicated is simply Gibbs free energy captured in bonds within the physicochemical molecules constituting the organism. The complexity of the communication system is the result of the selection of the most-energy-efficient forms during periods of energy scarcity.

When a forager ant discovers a new source of food, it lays a pheromone trail on its trip back to the colony. Other ants use this trail to locate the food, reinforcing the original trail with pheromones of their own. Soon, a steady line of ants is trekking from the colony to the food source and back again, laying down fresh pheromones all the while. They are drawn to the pheromones, like a slime mold to cAMP; the superorganism has extended an appendage. When the food finally runs out, ants stop producing the pheromones that reinforce the trail. As the trail dissipates, the ants cease to follow its course. Many small signals result in a concerted action.

Ants use chemicals for more than just finding their way around. To an ant, pheromones can mean the difference between war and peace. Chemical signals tell ants who is friend and who is foe, holding the colony together while helping it to identify and attack outsiders. When an ant is crushed, it releases pheromones from its body that send its comrades into a frenzy of aggression, and also attracts more ants into the melee. Pheromones are the local rules that produce the emergent effects of collective ant behavior, whether it is a picnic, raid, or a call to war.

10.7 Hijacking Communication Systems

In the 1890s, insect immigrants from Argentina snuck aboard a cargo ship headed for Louisiana. This small family of travelers spread, and now dominates the West Coast of the United States, having grown to a population of billions of individuals whose ranks stretch from San Diego to San Francisco. And North America is not alone—this ant family has spread to Africa, Asia, Australia, Europe, and islands across the globe. Since its arrival in the United States, the Argentine ant has vastly expanded its territory, often at the expense of native ant species. Because all Argentine ants in America are descendants of that single colony that landed in Louisiana, they all share pheromones that identify themselves to one another as friends. This prevents infighting and helps the invasive species spread unchecked.

However, recent research has discovered that a simple chemical change can disrupt the peace and harmony among the colonial domain of the Argentine ant. U.C. Irvine biologist Neil Tsutsui and his research team have found a way to turn these ants against each other. Tsutsui developed chemicals whose structures are similar, but not identical, to the pheromones that coat Argentine ant colonies. When a single ant was covered with the new chemical and placed in a petri dish with ten of its friends, it was immediately attacked, its limbs torn apart by the others. With these chemicals, the researchers hope to incite a civil war to halt the advance of this highly invasive species, whose depth of cooperation has formed a united front against its opposition.

This experiment shows how sensitive their communication system is to minor disruptions. A simple change in the pheromone's molecular structure is enough to turn cooperators into enemies. Spreading mismatched information, by confusing the ants' communication system and making them seem like enemies, radically changes the way in which individual ants interact. With such chemical alterations added to the mix, survival becomes less about who you are related to, as suggested by Hamilton and Dawkins, and more about the information conveyed. Because simple, local interactions are the foundation of the global emergent structure and determine the function of the superorganism, chemical tampering can cause a colony to devolve into warfare; the superorganism battles itself. Such a phenomenon is reminiscent of the spread of misinformation through propaganda, as a backdrop for war and cooperation among human societies.

Slavemaker ants employ the same changes in chemical communication in order to trick the kidnapped slave ants into dedicating themselves to the colony of another species. By raising the kidnapped ants as pupae, they integrate them within the slavemaker colony's communication system. They are using information to exploit the slaves' selfless dedication to the genes of its superorganism.

This is a clear example of altruism outside the bounds of the "altruism equations" elaborated by Hamilton. No web of genetic relatedness can explain the behavior of the enslaved ants. Instead, we have an instance where the signaling mechanisms that evolved in the context of familial relationships are hijacked by another group for its own advantage. The slavemaker ants receive reliable workers, but the enslaved ants get no fitness benefit in return: that is, they do not get to pass on any of their genetic material whatsoever. The relationship is reverse parasitic, in which the enslaved ants gain no direct or genetic benefit from their selfless dedication, while the slavemaker thrives and procreates. Kin selection, as described by Hamilton, may form the basic motivation behind altruistic action, but communication and the manipulation of information is a much stronger indicator of interest. In the broadest scope, information transfer within the network, in combination with energy efficiency, drive the emergence of what we see as cooperation, altruism, or antagonism, whether between or beyond kin.

10.8 Imprinting and Brood Parasitism

The phenomenon described above appears to be essentially the same phenomenon as imprinting, which was studied and popularized by Konrad Lorenz during the twentieth century. After hatching geese in an incubator, Lorenz noticed that the first moving object those geese laid eyes on, during a "critical period" of their development, would become imprinted onto the young geese. Essentially, a young goose would act as if the object were its mother, whether that object was a wire-frame, a puppet, or a picture of an adult goose. They are imprinted by the electrochemical signals elicited during that critical period of their development.

Imprinting is part of what makes "brood parasitism" possible. Many species of birds, such as the cuckoo, lay their eggs in the nests of birds belonging to other species. A "parasite" egg hatches, and the young chick is raised by its new "mother" as though it were truly the host bird's own child. There is a high cost to the mother in raising these usurping chicks because of the extra energy she spends traveling to gather food for this very hungry chick. Sometimes, the new chick will even kill the biological offspring of its adopted "mom" by using a sharp mandible hook that falls off a few days after hatching. The parasite's original mother will also push one egg out of the victim bird's nest, thereby maintaining a constant egg count, and creating a less competitive environment for her own hatchling. Her egg has evolved to appear nearly identical to those of the host mother.

A female cuckoo born in the nest of another species will imprint onto its new mother and, later in life, this imprinting leads it to lay its own egg in the nest of the species that had mistakenly raised it.[114] In this way, the brood parasites use specific signaling and species-recognition cues to specialize in one type of host.[115] Brood parasite birds are not so different from slavemaker ants: in both cases, the cooperative, parasitic relationships of these animals arise from the use of signals. The cuckoo is simply a slavemaker in reverse, infiltrating a nest, rather than taking victims from their original colony.

10.9 Superorganisms and Energy Efficiency

The metabolic rate of a colony of a hundred ants should be a hundred times the metabolic rate of an individual ant. But given Kleiber's law that the metabolic rate of an animal is proportional to its mass, raised to the power of ¾, we would expect to see a lower metabolic rate as we add more individuals to the colony. Experiments indeed show that the metabolic rate of an ant colony scales negatively with colony size.

Interestingly, the power law for ant colony metabolism versus size is ¾—the same scaling exponent for metabolism versus body-size among multicellular organisms.[116]

But, does scaling result from cooperation, communication, or from some other factor? When placed together, worker ants with a limited communication network maintained a higher metabolic rate, which scaled according to a power law of one: identical to single-celled eukaryotes. There was no metabolic advantage for this group of workers when they were separated from the whole colony. This demonstrates that the energy efficiency we see in the ant colony superorganism, which scales like a single multicellular organism, results from specialization and information exchange to create specialized ant groups performing specific tasks. Similar results were obtained from other studies on social insects, including bees and wasps.[117] [118]

Beyond insects, we find that other superorganisms scale the same way.[119] For example, groups of marine ascidians, simple colonial filter-feeding animals, show a ¾ metabolic scaling, but only when the individuals are connected in a network that allows for information transfer.[120] On their own, they scale following a power law of one—the same as the worker ants measured in isolation from the rest of the colony. We can see that increased communication, higher degrees of interconnectedness, and information transfer among specialized groups are at the root of energy efficiency in superorganisms, and these traits combine to give the superorganism higher genetic fitness. In the above study, growth rates in the largest colonies were up to seven times that of smaller colonies, and larger colony metabolisms were 30 percent more efficient.[121]

10.10 Evolutionary Success

Within the superorganism of any ant colony, we see the basic physical rules leading from chaos to complexity, which in turn leads to the emergence of surprising, unpredictable patterns of order. Given simple chemical communication, the interactions of the different ant castes result in the amazing, concerted actions discussed thus far.

Ant populations contain some of the most specialized individuals on the planet, so specialized as to give rise to a superorganism that has spread to nearly every surface of the globe. The more than 22,000 ant species make up an estimated one quarter of the terrestrial biomass.[122] [123] Their colonies can range from hundreds of ants to as many as 300 million workers and one million queens.[124] Ants are an extreme evolutionary success. Intense specialization makes the colony behave interdependently, and the superorganism that emerges from this interdependence is more energy efficient as a result. Effective communication increases energy efficiency, both in an ant colony and in any other superorganism. It is this increased efficiency that drives the evolutionary fitness of the species. In the next chapter, we will explore the workings of

Thermoinfocomplexity in social animals with greater individual independence.

Chapter 11 – Bonding and Social Networks

11.1 Social Network of Vampire Bats

Vampire bats are not picky when it comes to feeding. Night falls, and they go on the prowl for birds, mammals, and almost anything with a pulse. But the vampire bats do not just take blood, they also share it. Bat societies are built on the premise that everyone goes out at night to gather Gibbs free energy—blood from large animals, mostly—and, once they return, they share it with one another. Not every hunting bat is successful every night, so if the bats did not share, some would starve. To stave off hunger, a bat simply sidles up to another bat

243

that has had better luck that night, grooms it and licks its face. The signal is clear—I'm hungry—and the successful bat regurgitates a bit of blood for the luckless groomer. A vampire bat will often "donate" blood to its cave-mate, even when they are not genetically related.

The key to vampire bat cooperation is memory. Vampire bats need to remember which individuals did what over a long timescale. Memory introduces an element of time that has not been especially important to any of the complex, cooperative systems discussed thus far. Slime molds, ants, and the cellular components of our body all communicate via chemical signaling. When chemicals react, the changes they bring occur over the short-term. The larger "flockers" we have discussed, commun-icate with one another via more-complex means like vision, smell, and touch; their complex communication is responsible for their organization in space over the short-term. They have that in common with slime molds, ants, cells, and molecules.

However, long-term memory coupled with sociality represents a distinct jump in the complexity of life and matter. The ability of many complex systems to maintain efficiency over a long period of time has been selected for. Sociality is adaptive; the individual agents just have to keep track of each other.

Simple rules govern communication in long-term cooperation across mammals, and the emergent result of this communication results in energy efficiency for the group. These rules, governing cooperation over time, make the complex societies of social mammals possible, from human beings exchanging goods and services, to chimps, dolphins, whales, and vampire bats. By looking at how social animals interact with one another over the course of their lives, we will gain insight into one of matter's most profound organizational jumps from molecule to human: the jump that allowed for truly social communities, with large gains in energy efficiency.

11.2 Sharing Blood Makes Sense for the Group

As we said before, periodic scarcity of energy drives cooperation and makes it necessary for bats to cooperate over long periods of time. If a bat goes without blood for 60 hours, its metabolic processes will diminish to the point where the bat can no longer maintain the minimum body temperature necessary to stay alive. Every night, each bat in a cave needs to consume, on average, 50 to 100 percent of its body weight in blood. If a bat bites its victim in the wrong way, unintentionally causing pain, it gets shaken off. This failure is quite common for young, inexperienced bats. Considering the high energetic demands of their metabolism, it is a wonder that any of them make it to adulthood.[125] Only by spending calories (Gibbs free energy) efficiently, and sharing efficiently, do bats offset the dissipative energy of their own demanding bodies. The Gibbs free energy law is constantly at work; the input of energy into an open system through the intake of food into the body maintains homeostasis and slows down the dissipation of energy as per the Second Law of Thermodynamics, staving off the rate of increasing entropy.

Bats sharing blood are a clear case of mammalian reciprocal biological altruism. For any bat, less blood means less time until starvation, but the cost of donating is far less substantial than the benefit enjoyed by the hungry recipient. One study showed that by donating blood, a bat might drop from 110 to 95 percent of its pre-feeding body weight. Because of such a donation, it would reach the starvation point six hours sooner. Let us suppose the recipient bat has not eaten in two days, and in that time dropped to 80 percent of its weight compared to when it fed. The recipient bat would gain about 18 hours until the starvation point, compared with the six lost by the donor. This study shows mathematically what already makes intuitive sense: the metabolic benefit of additional food is far greater to a starving individual, than is the cost of giving up that same quantity of food to a well-fed bat.[126] Nonetheless, there is no discounting the energy loss to the donor bat.

The magnanimous vampire bats are not necessarily genetically related. The theory of kin selection predicts that related individuals will be more likely to cooperate. Although it is true that some instances of blood-sharing include a mother bat feeding her children, much of it takes place between unrelated brood-mates. Their altruism is part of a much more general phenomenon of reciprocal energy exchange remembered as information.

When a bat donates blood, the recipient bat will return the favor: the quintessential case of what biologist Robert Trivers termed reciprocal altruism. This form of altruism occurs regularly between non-relatives; it is an altruism based on reciprocity rather than genetic kinship, and it is responsible for much of human social behavior. Sociality finds evolutionary stability because it leads to more competitively energy efficient groups, making it a basis for group selection throughout the evolutionary process.

11.3 Energy of the Social Network: Balancing Between Plenty and Starvation

For reciprocation to work, bats must be able to recognize one another. Researchers have found that a mother bat and her young recognize each other by their distinctive calls. A young bat employs three types of calls: contact calls, separation calls, and recognition calls. A contact call consists of a low-volume chirp that facilitates bonding and occurs whenever there is physical contact between the mother and her offspring. Contact calls are often coupled with grooming, which is how bats become familiar with one another. A separation call is emitted by a young bat when it is no longer next to its mother. Once its mother hears the call, she responds with what are called *Stimmfühlungslauten*, or "true-feeling calls." In response, the young bat emits a recognition call, letting her know it has heard her. When a mother hears the separation call of her young, she goes to the young bat, gathers it close, tucks it under her wing, and licks it. In this mother-child relationship, we see the groundwork for social recognition among adult bats.[127] The mother-child bond is the first step in building a communication network that will have adaptive benefits for adult bat populations.

The adult bats in the cave have distinctive calls that function as information-rich signatures, allowing them to recognize one another. They emit their calls when grooming each other and when creating associations between touch, sound, and individual identity. Bats tend to groom themselves about ten times as often as they would groom other bats. The grooming that bats do for each other does not have a major effect on parasite levels, which suggests that the grooming of others serves a primarily social function. Grooming enables a bat to determine how recently another bat has fed, and bats are much more likely to share blood immediately after they have been groomed. As a bat grooms its neighbor—usually under the wing, near the stomach—it receives information about how swollen its neighbor's stomach is.[128] Here we see the interplay between selfishness and cooperation, each contributing to the energetic efficiency of the whole, thanks to an intricate web of communication.

Long-term cooperative relationships pay off metabolically. Pairs of unrelated vampire bats are known to develop a "buddy system," by which two bats regularly share food with one another.[129] No bat is an all-time donor, and no bat is an all-time mooch. Life is suspended between the availability of Gibbs free energy, and the flow of energy as per the Second Law of Thermodynamics. Bats share when they are able and borrow when they need to. The result is a balancing of the energy-level playing field; rather than be too full or starved, all the bats take in a modest amount of calories every day. Each bat receives a metabolic benefit and is able to maintain its own structure better. This sharing is adaptive, leading to fitter bats as well as a more persistent bat population.

Because vampire bats must make a down-payment of altruism before they see any returns on the investment, if the recipient does not repay its debt, it stands to reason that any one bat could cheat the system, receiving but never donating blood. Why would bats consistently hand out favors, if there are no guarantees that those favors will ever be returned?

11.4 Memory and Bonding

By remembering cheaters, bats may punish them. This simple feedback system contributes to the structure of bat societies. Memory, in a broader sense, is responsible for much of the organization in the natural world. The cells of our body have their own type of memory; when our immune system encounters a new bacteria or virus, it fights the foreign substance by producing antibodies to identify specific dangerous intruders. Produced by lymphocytes, antibodies have a structure that is chemically variable. The immune system "remembers" previous bacteria and viruses by forming a kind of flexible molecular tool on these structures that detects the chemical composition of the particular foreign agent. In the future, whenever that foreign substance enters the body, antibodies in the bloodstream can quickly identify and get to work on destroying it. This memory allows us to administer vaccines to children to boost the cellular memory banks of their immune systems, giving their bodies a strong defense against those particular pathogens in the future.

We do know that our memories are bound to our emotional lives. The amygdala, a component of memory storage and recall, is part of the limbic system, a group of brain structures that influence emotion and behavior. The interrelation of emotion and memory is particularly important for reciprocal altruism. Our memories of past interactions with specific individuals are inevitably wrapped up in our emotional impressions of them. The memory-emotion system is not perfect. Memories of concrete objects and events are inevitably colored by our feelings at the time of the experience. In most cases, this association helps us align events with the appropriate feelings, which can influence our decision-making process in a feedback loop. If someone has cheated us in the past, we remember them and associate them with nasty and unpleasant emotions, which tell us to avoid helping them again. On the other hand, if someone cooperates with us regularly, we develop a sense of trust. In either case, emotional impulses guide us in making decisions. We remember who we trust, and who may deceive us.

Memories make the punishment of bat cheaters possible. Biologist Gerald S. Wilkinson discovered not only that bats have memory, but that memory also allows them to make decisions about whom to feed and whom to ignore. In an experiment, Wilkinson took bats from two distinct populations, none of which were genetically related. In one group, however, the bats were strongly associated with one another by a history of sharing prior to their relocation into the experimental population. In the other population where there had been no previous association, no such sharing occurred. The individuals from the group of previously associated bats shared preferentially with one another because they remembered one another as distinct, recognizable individuals. They could "count on" each other to share blood and acted accordingly. Not only that, but a bat that received blood from another bat would be much more likely to reciprocate during future rounds of the experiment. Like a positive feedback loop, sharing fostered further sharing, but only when the bats were familiar with one another.

11.5 Bubble Net Fishing Among Whales: Learning, Remembering, and Saving Energy

One of the most elaborate instances of group hunting occurs among humpback whales. The technique, called "bubble net fishing," begins when one whale dives below a group of herring and blows a spiral of bubbles while swimming in a wide circle. The bubbles rise in a curtain and surround the fish, which panic. At precisely the right moment, the other whales join the first whale and emit cries from the bottom of the net of bubbles. The fish swim upward, and in a coordinated lunge, the whales follow, swimming through the bubble cylinder in pursuit of the fish. The fish scramble upward until they reach the surface. Some fish turn back and attempt to dart between the whales, which by now are rising in unison with mouths agape, jaws detached to make for a wider opening. For the fish, there is no escape. Suddenly, in a coordinated manner, all the whales turn their fins so that their white undersides flash upward at the oncoming fish. The sudden whiteness blinds and confuses the fish, which reverse course toward the surface again.

From a nearby boat, one can first see the ring of bubbles, followed by a few fish jumping, and more fevered splashing. Then, with a crash, all the whales surge out of the water—each taking a huge mouthful of trapped herring. The skin of a humpback whale's lower jaw is pleated and able to expand into a pouch known as a "buccal cavity." As the whale hurdles upward, this buccal cavity fills, and then, using their baleen, the whales strain away excess water and take a big swallow of fish. The process then begins anew, each whale repeating its own unique role, every time.

This group-hunting technique is highly adaptive, and not all humpbacks engage in bubble fishing. Over the course of humpback whale evolution, bubble net feeders began to replace the whales that did not use this efficient feeding technique. In bubble net fishing, the role of each hunter is terrifically specialized. The bubbles are always blown clockwise, the vocalizations are perfectly timed, and each whale takes up the same position relative to the other whales, as they rise together through the curtain of bubbles. This sort of coordination typically occurs with groups of whales ranging from five to eight individuals, but there have been reports of up to 25 whales working together. Sociality bestows energetic benefits on each individual that simply would not be possible without such long-term coordinated teamwork. This group behavior is selected for because it enables each whale to meet the demands of effective energy capture. The input of energy into the whales' "open system" occurs at a lower energy cost than hunting individually.

This fishing technique is possible because of humpback whales' intricate sociality, which in turn is possible only because of their highly developed brains. This latter quality distinguishes whale group hunting from the group hunting of herrings. Herrings have evolved precise behaviors of cooperative hunting, but memory and learning do not come into play in the same way as they do in whales, where social complexity has led to increased fitness for the species. Energy efficiency in whales comes not only from tight intuitive coordination, as in herrings, but from practice and learning, a form of an exchange of information that extends far back in time. It works through a network dependent on memory and long-term cooperation.

11.6 The High Society of Dolphins

Dolphins are also extremely social creatures, nearly as sophisticated as humans. They reciprocate like vampire bats, and they engage in group-hunting like whales and wolves. With dolphins, we see the precise fine-tuning of the group dynamics, allowing for an emergent jump in sociality. Dolphins travel together in pods wherein related and associated individuals have unique whistles that allow them to recognize one another, much as with the calls we saw when we looked at bats. In studies where scientists played whistles back to subject dolphins, the whistles of the closest relatives and most familiar individuals elicited the strongest behavioral responses. These sounds make for some of the loudest communication in the open ocean, laying a vast network of information exchange.[130]

A study of female dolphins showed that females cooperating in pods have more and healthier offspring than lone dolphins.[131] Together, the females are able to better protect their young from sharks. Adult females help to care for the offspring of other adult females, like babysitters. They help each other rear calves, a tendency that further shapes dolphin social structure. Experienced mothers tend to associate most strongly with other females who also have calves, and by rearing young together, the adults teach the baby dolphins the techniques that will enable them to thrive once they reach maturity.

Dolphins also have an incredible capacity for learning. Much of their behavior is transmitted from parent to offspring. Dolphins are one of the rare animals that practice "cultural transmission." A culturally learned behavior is defined by Whiten, et al., as behavior that is "transmitted repeatedly through social or observational learning to become a population-level characteristic."[132] Because of their large social groups, dolphin learning takes place not only from mother to offspring, but from many adults to many offspring. Cultural complexity increases as the number of teachers and learners increases. This sort of interaction has led to at least one odd, learned dolphin behavior. Certain types of dolphins have learned to use sponges, broken from reefs, as a cap on the end of their noses, which they use as a tool to protect their snouts while gathering food.[133] Other culturally transmitted behaviors include "fish-whacking" and "kerplunking," by which a dolphin uses its fluke to stun prey, or flush prey out from hiding.[134] [135] Sociality, as well as a highly developed brain, makes learning possible. Sociality extends the lifespans of the individuals and ensures homeostasis at the macrostate level of the group. As we have indicated before, homeostasis results from a balance between energy intake (Gibbs free energy) and energy dissipation in entropy (Second Law of Thermodynamics). A more-efficient balance results in a higher fitness for the organism.

In dolphins, long-term cooperation between males also increases fitness. Bonding in male dolphins lasts for years and contributes to the success of each individual. Male dolphins pair off and cooperate in most of their daily routines. For instance, highly cooperative feeding techniques enable paired male dolphins to gather more energy per dolphin than either one could individually.[136] They eat better by cooperating, and pair-bonded males protect each other from sharks. Cooperation between paired males also directly enhances their reproductive success. In a phenomenon known as "mate guarding," two adult males will flank a female when she is reproductively receptive.[137] By flanking the female, the males prevent rival males from taking advantage of the mating opportunity. The result is that pair-bonded males sire more offspring than lone ones. As male dolphins develop a pair bonded relationship, their whistles begin to sound increasingly alike. This similarity in intonation among individuals that pair is shared by dolphins, chimpanzees, some birds, and even humans. Their calls indicate their growing familiarity and the increased efficiency of the information network.

Gender differences aside, all dolphins benefit from group-hunting techniques. Teams swim in circles around schools of fish, gathering them into what is known as a bait ball. Once the ball is formed, the dolphins take turns plunging through the center of it, nabbing as many fish as possible. Sharks get in on the action too, while opportunistic seabirds dive under the ocean surface to catch a choice meal. Another coordinated hunting technique includes corralling, whereby dolphins chase fish into shallow water, making them much easier to catch. Strand-feeding is another technique, practiced by the Atlantic bottlenose dolphin of South Carolina, in which the dolphins drive fish not just into shallow water, but onto mud banks at the water's edge, where the fish are especially vulnerable.[138] All of these cooperative social behaviors contribute to the energy efficiency of the dolphin group, increasing its evolutionary fitness.

We have seen bats, whales, and many other organisms benefit from the emergence of sociality, but dolphins take sociality to the next level. Their cooperation is not limited to what can be explained by kin selection or reciprocal altruism. Dolphins often act altruistically in situations without guarantee of reciprocation, and among those of no genetic relation. Viewed in this light, their cooperation is a much more general phenomenon. Among dolphins, we find an example of pure altruism. There have been reports of dolphins ushering lost sperm whales that have wandered into fresh water, back to the ocean.[139] This altruistic act has no known benefits for the dolphins, short-term or long-term. Countless surfers have been attacked by sharks, and then rescued by small teams of dolphins swimming in defensive barriers until the sharks leave the area.[140] Perhaps blood signals them to huddle together in self-protection, we may speculate.

In Brazil, one example of regular cooperation between humans and dolphins is particularly well recognized. Dolphins herd fish to a shoreline where humans wait, ready to cast nets. The dolphins give a signal as part of an intricately ritualized interaction with the human fishermen, a signal that tells the fishermen precisely when to throw their nets.[141] The fishermen haul in their nets, full of fish and the dolphins catch many of those that try to dart away.[142] This may simply be an example of fisherman utilizing dolphins fishing strategy to catch more fish.

11.7 Chemistry of Social Bonding

Social bonding can be a profoundly general phenomenon. In Nashville, Tennessee, for instance, an Asian elephant and a dog living in a sanctuary made headlines in 2009 for being "pair-bonded." They played together, ate together, and spent their days together. When the dog, Bella, had to stay indoors for a while because of an injury, Tara, the elephant, waited outside for the duration of the recovery.

Bonding is not always based on genes, reciprocity, kin, species affiliation, or symbiosis; it is often based on much more concrete mechanisms, like the electrochemical signaling within and between the bonding individuals. In mammals, several hormones lie at the heart of social bonding—from mother and child, to mating partners and companions.

The first hormone to be identified, oxytocin, was discovered in 1909, when Henry Dale found that an extract of the pituitary gland caused uterine contractions. He derived the name "oxytocin" from the Greek word for "quick birth," and later found that oxytocin played a role in the contraction of muscles around mammary glands, causing the milk letdown phenomenon. For many years, scientists thought the role of oxytocin was limited to childbirth and breastfeeding. Kerstin Uvnas Moberg, a Swedish scientist, discovered the connection between oxytocin and positive emotions. He termed the oxytocin response the "calm and connection system," because of its dual action on the body and the mind.[143]

However, oxytocin is not solely responsible for social bonding. Human brains (like those of other mammals) are bathed in a complex cocktail of chemicals that control their functions. These neurotransmitters trigger fast responses and influence the responsiveness of brain cells to various chemicals.[144] In humans, the molecules responsible for emotion include arginine vasopressin (closely related to oxytocin), serotonin (important in mood regulation), norepinephrine, (a molecule triggering excitement), dopamine (the molecule of pleasure), and estrogen and testosterone (which control the reproductive system and regulate processes of the brain, including mood and emotion). When neurotransmitters are released, they attach to protein receptors on the membranes of nearby cells, causing a chemical cascade that leads to the appropriate emotions. Hormones work in a similar manner, though they travel through the bloodstream first to reach the appropriate receptor-cells and tend to cause longer-lasting responses. While oxytocin is found in many mammals, including bats, oxytocin-like neurochemicals are found in other animal groups as well: isotocin in fish, mesotocin in lungfish and some tetrapods, and vasotocin and mesotocin in birds.[145] In different species of finches, mesotocin receptor distribution in the brain correlates with flock size. The administration of mesotocin increases social behavior, such as flock formation, while mesotocin antagonists reduce such behavior.[146] In another example, it was shown that oxytocin and arginine vasopressin are instrumental in social interactions of the prairie vole (*Microtus ochrogaster*), and an absence of vole social interactions was observed when the concentration of these neurotransmitters drops below a certain level.

The body's stress response circuit is known as the hypothalamus-pituitary-adrenal (HPA) axis. Its command center is the hypothalamus, a part of the brain that manufactures many of the "molecules of emotion." The hypothalamus is not only the center of the HPA axis, but also the switchboard for the oxytocin loop. It produces oxytocin and releases it directly into the central nervous system. It also sends oxytocin to the pituitary gland, which stores it and then releases it into the bloodstream in response to certain stimuli. The end-point for the oxytocin system is receptors located in the brain, as well as, in the rest of the body. While there are oxytocin receptors all over the body, the one that produces feelings of affection and connection are located in the parts of the brain that handle interpersonal relationships. These brain circuits are sometimes known as the "social brain," and oxytocin makes the social brain circuits come alive.[147]

In humans, a hug or any loving touch can produce a sudden increase in oxytocin. When oxytocin reaches the brain's social center, it interacts with other feel-good chemicals to make us relate intimately with a particular person, creating the bond we call love, whether it is maternal behavior, marriage, or the trust we place in a friend.[148]

11.8 Complex Brain, Complex Social Network

The tendency to bond and, in many social mammals, cooperate is pervasive throughout the natural world. Because of this tendency, an elephant and a dog can act as if they were born in the same litter. The bonding ability may have arisen by evolutionary selection, but as in the case of Tara the elephant, it is clearly the result of electrochemical information exchange reinforced by memory. It is extended beyond the usual boundary of social bonding that led to evolutionary fitness.

The complexity of the brains of higher animals makes it possible to remember and learn. As we saw in this chapter, it led to the next level of social network complexity that is possible for dolphins but not ants. Complex social networks, enhanced by bonding and cooperation, fuel emergent qualities that make groups more energy efficient, adaptive, and resilient to fluctuations in available energy present in the environment. In response to changes in energy availability in the environment, complexity builds on complexity, efficiency builds on efficiency, and, as with any other emergent system, there is a transition point where the interacting parts of a system suddenly give rise to a new emergent pattern at a certain threshold. This threshold, at the critical point for emergent systems, is directly related to the complexity of the system, the interconnectivity of the agents within the network, and the frequency of their interactions. Just as Moore's Law states that the capacity of computer chips tends to double every two years, biological evolution builds upon the previous level of complexity, albeit along a larger time frame. Greater complexity coincides with gains in energy efficiency. The next two chapters deal with human society: the most complex information network on Earth.

Chapter 12 – The Human Superorganism

Viewed globally, the human superorganism is composed of societies that can be clustered into anthropological and sociological categories such as hunter-gatherer, agrarian state, and industrial nation. These categories, while useful shorthand for historical and evolutionary narratives, hinder attempts to explain the dynamics of the evolutionary process.[149] Viewed from the bottom-up, every society is composed of individual human organisms seeking to thrive by interacting with one another through energy and information exchange networks. Just as every multicellular organism is a natural hierarchy of eukaryotic cell networks clustered into organ systems, every human society is a network of individual human beings clustered into families, villages, cities, and states.[150] As we shall see, the complexity of human societies is a function of

264

both the environmental sources of Gibbs free energy available to sustain the society, and the efficiency with which the society transforms this available energy to create and maintain its structure.

Here we must note that human societies do not march uniformly along a path of progress toward "higher" levels of complexity. Social complexity emerges in proportion to population pressure and the intensity of energy and information interactions between constituent subsystems.[151] [152] [153] The human societies emerge from the coordinated activity of self-similar individual adaptations to specific environments. Each regional environment in the global ecosystem exhibits differential selective pressures on the societies that inhabit it.[154] As such, features of human societies are selected for on the basis of their ability to thrive in their local environment.[155] Viewed at the broadest scale, the complex energy and information exchanges both within and between societies and their environments constitute the metabolic and developmental activities of the human superorganism, which has grown to inhabit nearly every region of the planet.

12.1 Scarcity Breeds Complexity

In Chapter Seven, we introduced the slime mold *D. discoideum* as the paradigmatic example of how complexity emerges in response to energy scarcity. There we examined how the slime mold undergoes a phase transition from a population of unicellular organisms to a multicellular organism. This process occurs through the autocatalytic diffusion of cAMP, triggered during periods of starvation. The slime mold demonstrates the capacity of living systems to maximize metabolic energy efficiency, not only through the scaling of body size as discussed in Chapter Eight, but also through efficient communal distribution of nutrients from the environment.

In a remarkable experiment, a team of Japanese researchers demonstrated the power of slime mold as a model organism for the emergence of complexity. When placed in a labyrinth with two different food sources located at different points in a maze, the slime mold connected the two food sources with its extended plasmodium. By connecting the two food sources, the slime mold distributed the food throughout the network of its constituent cells. The manner in which the slime mold accomplishes this feat is remarkable. The organism finds the shortest path between the two points, as shown in *Figure 12.1*. Initially, the slime mold will explore all possible paths through the labyrinth. The pseudopodia that fail to find food will die off, shrinking back into the main body of the plasmodium. The shape that the slime mold's "body" ultimately takes through the labyrinth is exactly the most-efficient metabolic solution to distributing Gibbs free energy throughout the complex adaptive system.[156]

Here we have a prime example of energy efficiency, as the driver of natural selection. The slime mold maximizes its foraging efficiency, and therefore its chances of survival, by finding the shortest path between its food sources.[157] Evolution's tendency to minimize differences in environmental Gibbs free energy concentrations, results in the most-efficient distribution of free energy throughout the cellular network. Through entrained processes of contraction and expansion, the slime mold finds the shortest path between the two food sources, sharing information about the location of the food sources throughout the network, and capturing that information through the processes of intercellular signaling. The final shape of the slime mold's plasmodium is a maximally efficient network of energy and information exchange, exhibiting the intelligence inherent in nature at all levels.[158]

Figure 12.1 Slime mold solving a labyrinth. The left image depicts the slime mold exploring all possible avenues through the maze. The right image depicts the slime mold's solution to

267

the maze, which is the shortest distance between the food sources.[159]

The reader may wonder what relevance an extended discussion of the slime mold has to the evolution of human societies. Throughout this book, we've made the argument that new complexities as they arise are selected because they are energy efficient in the face of energy scarcity. In the case of the slime mold, we've seen how aggregation and signaling between individual cells increases the fitness of the overall group, providing the emergent fruiting bodies and spores access to new food sources that were unavailable to its constituent cells acting on their own. For biologists, the slime mold is a model organism for the study of the evolution of social cooperation, as it demonstrates social cooperation in their simplest forms.[160] The slime mold is a microscopic version of the macroscopic phenomenon of complex organisms forming a superorganism, whether the population is of ants, bats, or human beings. As we shall see, much of the structure of human societies is a function of the need of every human being to secure a living from their environment. As such, the evolution of complex human societies is yet another example of the evolution of complexity that we have discussed in the previous chapters.

12.2 Bonding Energy: Human Societies

Human ecology is the study of the complex system of interrelationships between individuals, societies, and their environments, which characterize the evolution of the human superorganism. At the level of the individual organism, human beings are quite metabolically efficient. Human beings are not only one of the most metabolically efficient organisms but are also particularly slow-growing mammals. However, their average daily metabolic need requires more food than can be obtained by a single foraging individual. Thus, cooperative foraging and food sharing are emergent behaviors essential to the survival of the human species.[161]

It is impossible to discuss the evolutionary fitness of humans without acknowledging the critical component of the social structures in which they live. Unlike most of the social organisms we have studied this far, human beings have developed behavioral patterns of hunting, foraging, and sharing food that are not purely instinctual; they must be learned through culture. Thus, human evolutionary fitness is as much a function of stable sociological networks, as it is of robust biological processes. From the Thermoinfocomplexity point of view, the distinction between "biological" and "sociological" is immaterial, yet the sociological structures that characterize the human superorganism exhibit additional complexity that must be examined at each additional level of emergent complexity of various social structure.

In terms of their metabolic needs, human societies are no different from any other group of organisms. Like the flocks of birds, schools of fish, packs of wolves, and colonies of ants

269

discussed in previous chapters, human beings self-organize into societies that maximize energy efficiency for the entire group. As we have seen, complex adaptive systems evolve in ways that maximize their energy efficiency, capturing Gibbs free energy from the environment into bonds of energy and information networks in the process of evolving maximally efficient metabolic social structures. In this chapter, we will observe how these networks exhibit self-similar fractal scaling through the social bonding, migration, and settlement patterns of human societies. The evolution of human societies reflects the same evolutionary processes we have observed from atoms to organisms.

Despite the diversity of cultural forms, we observe that every society organizes itself in a nested hierarchy of relationships. Since no human being can fulfill his or her daily metabolic needs in isolation, it is natural to expect that patterns of Gibbs free energy distribution can give us clues to understand the emergence of complexity in different human societies. Cultural networks naturally emerge as complex adaptive systems self-organizing through the sharing and distribution of food, energy, and information between people in a society.

In every human society, the highest-frequency exchanges occur between people clustered into some form of nuclear family, composed of parents, offspring, and their closely related kin.[162] Throughout human societies, we find nested hierarchical relationships echoing those found at the scales of complexity already surveyed.

A version of the "first cousins" family group constitutes the first hierarchical extension of the social network from the microstate of an individual family (composed of a "nuclear" family of parents and their offspring) to a macrostate composed of a single set of grandparents, children and their spouses, and grandchildren. This family group, in turn, is a social network that maintains ties to the family group of descendants of the grandparents' siblings. Basic family architecture closely reflects genealogical family trees when organized by time, but from an energetic perspective, the patterns of family bonding cluster into societies that appear as self-similar fractal networks (see *Figure 12.2*). Variations of this clustered family network form the basic social architecture of societies at every level of complex organization.

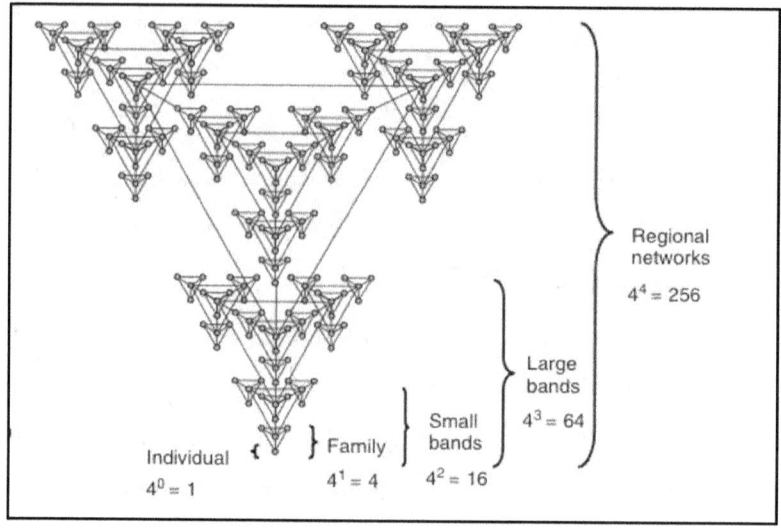

Figure 12.2 Diagram of self-similar topology of human social networks. Social groups are hierarchically structured with a scale invariant network topology.[163]

An individual human being can only manage so many close personal relationships. The bonding strength of these relationships is proportional to frequency of interaction and how much energy and information is exchanged through social interactions. The strongest bonds are formed between closely related kin, each of whom has a dynamic set of relationships, maintained over the course of a life. Beyond kinship ties, people maintain weaker bonds with people with whom they communicate less frequently, and these more distant relationships are the bonds that hold the greater society together.

The nested pattern of microstate-macrostate emergence self-organizes as human individuals cooperate to meet individual metabolic needs through the cultural exchange of energy and information. Social bonds self-organize across hierarchical scales of network organization, creating social "bodies" that increase the efficiency with which individuals in the society meet their metabolic needs. This metabolic efficiency scales in a similar manner to the metabolic scaling of organisms that we have discussed in previous chapters.

The degree of social network overlap between people is a measure of their relatedness, and the extent of one's social network developed over the course of a lifetime. For example, as children, brothers, and sisters will share social networks that overlap with one another almost completely. As people mature, their social networks extend beyond the immediate family relationship into the larger social network. Though there will always be significant overlap, social networks of siblings will diverge over the course of a lifetime. Each person will develop his or her own unique social network of micro-macrostate energy and information exchanges that ultimately extends throughout the society. Families repeat the network pattern at a higher level, combining microstates within a larger social network macrostate. Taken together, the sum of this interactive bonding between human beings forms an extended social network known as a society. These self-similar fractal patterns of complex organization are repeated across vastly different cultural and environmental conditions.[164] [165]

12.3 Foraging Systems

For most of human history, human societies have self-organized into family-level groups, whose structures are primarily shaped by the pursuit of maximally efficient foraging strategies.[166] Foraging societies have been studied extensively by anthropologists, who have gathered vast quantities of ethnographic data that can be examined across cultures. From an energetic perspective, these societies have adopted remarkably similar patterns of social organization. Overall, the population densities of hunter-gatherer social networks are quite low, and over the course of the year are a function of changing environmental conditions and concentrations of available resources.[167][168]

Foraging societies self-organize into fluid and mobile family-level groups composed of small bands of four to five families. Groups of families maintain social bonds with others, extending the network into larger bands of four to five family camps. These bands maintain social ties with extended regional networks that can range from many hundreds to thousands of individuals.[169][170][171] As might be expected, families cluster together into more highly structured social groups when resources are scarce and locally concentrated, and break apart into smaller groups when resources are more plentiful and widely distributed. Though the particular form varies, this cyclical pattern of concentration and dispersion in response to available resources is present in every human society. The organization of the society changes over time in order to optimize energy efficiency in response to fluctuations in available environmental energy. Throughout this chapter, we will see that the quantity and predictability of the water supply is a primary factor in determining the availability of environmental free energy, as well as a determining factor in the complexity of human societies.

12.4 Balancing Wet and Dry: The !Kung

One of the most extensively studied foraging societies is the !Kung of Africa's Kalahari desert. They inhabit a vast area, covering roughly 350,000 square miles on the southern tip of the African continent.[172] The Kalahari has a dry and cool winter, followed by a hot and rainy summer. Resources are erratically distributed, with the most-consistently available food sources clustered around the more-permanent water sources. Water is a limited resource, with few permanent water sources as well as a high variability in rainfall from year to year. Such limited resources will not support large populations. But, at low population densities, there is an abundance of available food sources. The !Kung have adapted to life in this landscape, maintaining a broad diet by optimizing adaptive strategies to maximize their access to seasonally available food resources while minimizing the effort required to do so.

A typical !Kung campsite consists of roughly five huts, each housing a nuclear family. This network of five families will "eat its way out" from their campsite over a period of a few weeks, moving on when the locally available food supply has been depleted. The longer a group inhabits a particular location, the further out it must travel from its base camp in order to procure food. When the local resource supply becomes depleted to such a degree that the effort required to gather food from that campsite becomes more than the effort required to move the camp to a new location, the camp will move. The foraging group will then repeat the process at a new campsite, traversing the entire circuit over the course of the seasons, as they follow the fluctuations in availability of environmental resources.[173]

Group populations alternate between periods of concentration and dispersion in response to available energy density. In the dry winter, extended regional kinship networks congregate into camps clustered around the few permanent water sources. When the spring rains come, the renewed energy abundance enables the large networks to break into the smaller family units, which disperse throughout the region. From these camps, they venture out on daily foraging trips until they have "eaten their way out" of the base-camp, and then move on to another location. Over the course of the summer, they repeat the wider yearly pattern of congregation and dispersal on smaller scales, continuing through the seasonal round until the larger regional networks congregate again around the permanent water sources during the dry season.[174]

During the period of abundance, the smaller groups disperse widely in order to maximize the inter-group distance, ensuring the most-efficient resource consumption for all groups. When food becomes scarce, the smaller groups will again converge in larger regional networks. This necessarily complicates inter-group relational dynamics as people self-organize into larger groups, exhibiting increased organizational complexity in the face of energy scarcity. Examined through a thermodynamic lens, the movement of !Kung foraging groups displays a metabolically efficient pattern of resource consumption. Though the circuit never repeats itself exactly, the overall pattern of annual "orbits" through the landscape is an attractor pathway. The simple process of !Kung families seeking to feed themselves over the course of the seasons, constantly moves toward the maximally efficient circuit through their environment. This minimizes the work required to consume the available Gibbs free energy captured annually by the plants and animals that cohabitate the Kalahari landscape.

Figure 12.3 displays an actual map of !Kung foraging trips. As they migrate in response to the seasonal fluctuations in the distribution of food resources in their environment, the !Kung follow the seasonal availability of water in the arid landscape.[175] These seasonal fluctuations in available energy density, captured and made available by plant and animals following the same fluctuations, provide the attractor pathways traveled by the !Kung throughout the seasons. *Figure 12.3* is a

277

vivid example of the complexity dynamics that undergird the formation of human societies. Indeed, this annual pattern of !Kung foraging along these attractor pathways exhibits a maximization of energy efficiency through time, analogous to the spatial propagation of the slime mold noted earlier. Here, again, we find the self-similar pattern of maximally efficient metabolic organization observable at all scales, operating at the level of human societies.

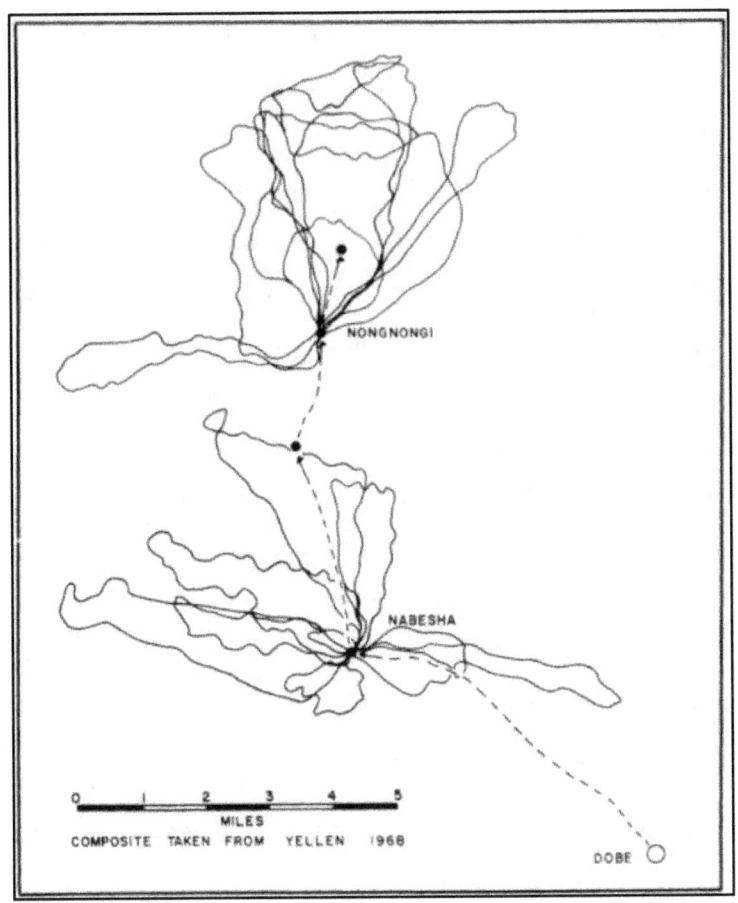

Figure 12.3 Map of !Kung foraging trips. Note the movement of
base camps and radial movement of resource procurement.[176]

12.5 The Bonding Structure of the Human Superorganism

When we examine society as a metabolic
superorganism, we find many of the same scaling phenomena
that we have noted throughout this book.[177] [178] [179] In ecological
networks, the rate of a population's resource consumption scales

with population size in a manner similar to the scaling of metabolic rates observed in multicellular organisms.

A scaling law in ecology notes the relationship between the size of an organism's home range, and the rate of resource use required by that organism. The home range of the organism is an area defined by the rate of resource use divided by the available food in the area. Larger organisms will require greater foraging areas, but like metabolism, the area used by groups of organisms does not necessarily scale linearly with increasing numbers of organisms; it is, rather, characterized by a scaling exponent, β. By now, we should expect that the greater the complexity of the system, the more efficient the scaling should become; and indeed this is what we find with human populations. The area, A, required by a population of N individuals to meet their metabolic requirements is $A = HN^{\beta}$, where H is the home range of an individual. If $\beta = 1$, the area required by the population, is the sum of the areas required by each individual. Moreover, the required area scales sub-linearly, meaning that the population uses space more efficiently as it increases in size. The space required by an individual to meet his or her metabolic needs decreases with increasing population size.[180] [181]

When the !Kung and other foraging groups congregate into larger, more complex regional social networks, during periods of energy scarcity, the overall metabolic efficiency of the group increases. During the dry winter, the more-complex social system uses the available food more efficiently than do the more widely distributed family-level groups during the abundant summer. However, it is important to note that even though the regional network is more efficient in its resource use per individual, it also consumes more resources per unit area while the group is congregated.[182] This inevitably leads to social tension when resource supplies become depleted; thus, the larger regional networks break apart and disperse into smaller family-level units, in time to seek the abundant resources of the coming spring. Later in this chapter we will examine what happens when such dispersal is not an option.

As in atoms, molecules, and other organisms, complexity naturally emerges as human populations congregate in the face of energy scarcity. Subgroups of families nest self-similarly into fractal social networks of increasing complexity, gaining energy efficiency as the adaptive system integrates additional levels of organizational complexity.[183] Like all complex adaptive systems, this process of self-organization captures available free energy into social networks of material, information, and energy exchange.

The primary characteristic that distinguishes foraging societies from more-complex and hierarchical human social networks is the extent to which foragers rely on social ties to minimize risk in the face of energy scarcity, rather than on the physical storage of food.[184] At all levels of the superorganism of human society, the ties of reciprocity are to individual human

281

beings what covalent bonds are to atoms in molecular compounds. Molecular communication is to bacteria in quorum sensing, as ties of reciprocal altruism are to bat colonies. Human beings have developed a deeply ingrained sense of mutual understanding that values relationships, as well as the deep level of trust that emerges between people who frequently exchange energy in the form of gifts and other goods and services.

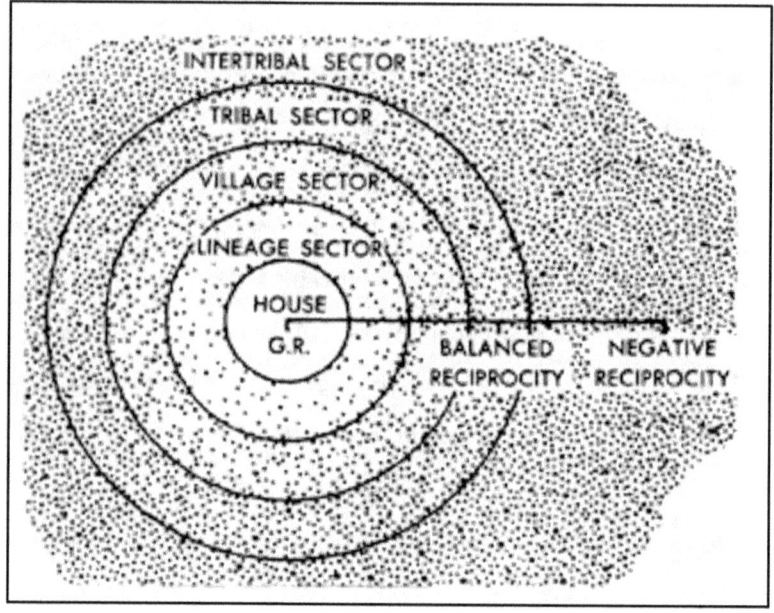

Figure 12.4 Attractor basin of reciprocity as a function of social distance.[185]

12.6 Basins of Attraction

In human societies, the basic family unit is defined by the underlying assumption that a free flow of reciprocal exchange will occur between the people that compose the unit.[186] Within a family unit, there exists a generalized obligation of reciprocity, which is the primary cohesive force that binds a nuclear family together. Indeed, this sphere of general reciprocity is the point attractor around which the complex system of the family orbits (see *Figure 12.4*). Many entrained behavioral patterns self-organize around this attractor basin, and the iterated multilevel stacking occurs through microstate-macrostate exchange networks, which form the general architecture of a society. At this level, Gibbs free energy from the environment is captured into the social bonds that form the very fabric of human society. As we shall see, this basic network architecture scales beyond foraging societies to encompass human social networks at every level of complexity.

In foraging societies, complementary entrained behaviors emerge through the division of labor within the family, on the basis of age and gender. Gendered division of labor, like the complementarity between electric charges, chemical affinities, and so many other polarities across scales of complexity, increases the efficiency with which Gibbs free energy from the surrounding environment is captured into bonds maintaining the organizational structure of a complex system. The formalization of generational and gender expectations within families creates sustained bonds of mutual interdependence.[187] Within this sphere of general reciprocity, energy and information resources flow so freely that it is impractical to keep a strict accounting of who owes what to whom.

Beyond the "atomic" sphere of generalized reciprocity within the immediate family, the social network becomes more complex as methods of accounting emerge in the extended kin group consisting of cousins, uncles, aunts, and grandparents—though grandparents often live within the immediate family sphere. In this sphere, there is a general expectation of balanced reciprocity, and a strict accounting of the fairness in exchanges over time.[188] Non-givers are brought into compliance through embarrassment or other social mores, and will ultimately be ostracized from the social network, if they fail to comply with the general rules of reciprocal exchange.

Beyond this sphere of consanguine ties, lies the larger social network, which in foraging societies, congregates during periods of energy scarcity. Here, it is important to note that rather than defining relationships in a top-down manner, the social network itself is created in a bottom-up process through the activity of energy and information exchange networks. These networks form the very fabric of the society. Foraging societies self-organize through processes of reciprocal exchange that connect individuals into families, families into bands of related kin, and kin groups into villages and regional networks.[189] [190] In order to procure a living by foraging, a person must be free to pass through the territories where he or she gathers food. The tribal network is also the level at which more-abstract bonding between groups takes place, through ceremonies, feasts, and celebrations.[191] Individuals meet their spouses at gatherings, consecrate unions through marriage rites and feasts, and share stories of bravery and camaraderie recorded in tribal lore.

Methods of obtaining Gibbs free energy from the environment, capturing this energy by sharing it with other people, and sharing culturally encoded information about these resources through bonds of social relationships are the cultural processes that constitute the basis of human societies. They represent the processes of energy capture in the bonds that build and maintain the complex networks that we have observed at every level of the Thermoinfocomplexity hierarchy, from molecules to human societies.

As we have observed throughout this book, opposing forces of competition and cooperation also operate at all levels of the human social hierarchy. Competitive forces inevitably emerge through the need of every individual to obtain food from the environment. In foraging societies, the competitive forces fragment the complex social structure into smaller groups that diverge from one another in order to more efficiently gather food during times of plenty. Knowing that periods of energy scarcity will return, groups are drawn together by cooperative forces for feasts of sharing, celebrating their successful efforts during seasons of bounty. The widespread harvest rituals observed by human cultures reinforce the reciprocal exchange networks that will hold the groups together in order to survive inevitable periods of scarcity.

The constant push-pull between these forces of competition and cooperation scale-up through human societies of every level of complex organization, constituting the basic sociobiological structure of the human superorganism. The "surfaces" of these spheres, organized around the attractor systems outlined above, become social membranes that enable microsystems to nest within macrosystems, which in turn become microsystems for higher macrosystems across every level of a given society. Beyond the largest membrane of shared social identity, lie the "other" neighboring groups whose sustenance is seen as competitive with the primary group.

The structures outlined above work to maintain human societies so long as internal population pressure is low, and neighboring populations have non-overlapping territories from which to gather subsistence from the environment.[192] [193] When population or environmental pressures force multiple foraging societies to compete with one another, conflict ensues. Beyond the potential for conflict, however, lies the potential for different attractors to emerge, ones that can organize the forces of competition and cooperation into social networks of greater complexity.

12.7 Agriculture's Deep Origins

The origins of agriculture have been traced to about ten independent centers of origination, all occurring between roughly 11,000 and 5,000 years ago.[194] The earliest origin occurred in the Fertile Crescent of the Near East, 11,500 years ago. Northern China saw the next independent emergence a few thousand years later, and the most recent development occurred in Central and South America. Although diverse local conditions governed the emergence of each agricultural society, all agriculture evolved after the end of Pleistocene Era, as the climate shifted toward the more climatically stable Holocene period.[195] [196] [197] [198] [199] As we have observed in the case of Hawaii, agriculture first emerged in stable climates in which populations grew to reach the environment's carrying capacity. In this section, we will explore the evolution of complexity in the human superorganism under geographic conditions with fewer constraints than those of the Hawaiian archipelago.

An exhaustive history of the origins of agriculture is beyond the scope of this book. To further demonstrate the applicability of Thermoinfocomplexity theory to human societies, we note that everywhere agriculture emerged, attractor pathways, i.e., the paths of least resistance for energy flow, dictated the timing as well as the precise locations of agricultural projects. Once established, agricultural communities accumulated surplus energy in positive feedback loops in accordance with the dynamics of the autocatalytic set of social

287

hierarchies, labor specialization, and land tenure systems. The increased heat capacity, or cumulative total of Gibbs free energy bound up in social structures, exerted competitive pressure on nearby populations. Just as heat flows along a thermal gradient from hot to cold, the thermodynamic energy flow between more energy dense agricultural societies and their "surroundings," populated by foraging societies organized at the family-level in a lower internal energy density, resulted in a competitive ratchet favoring the spread and dominance of agricultural civilizations.[200]

12.8 Agriculture and Earth's Exponential Population Growth

The success of agricultural societies is evident. Between 70,000 B.C. and the origins of agriculture 10,000 years ago, the Earth's human population was stable, at around one million people.[201] Agricultural societies dramatically increased the efficiency of food production, increasing regional carrying capacities for human populations by channeling environmental production of Gibbs free energy into human societies. The rapid growth of the human population in the wake of the agricultural revolution is a testament to the effectiveness of agricultural practices in sustaining the human superorganism. For example, from 300 to 400 A.D., the Roman Empire consisted of between 50 and 60 million people.[202] By 1340, the population of Europe had grown to over 70 million.[203] Global population further boomed in the 18th century with the Industrial Revolution. The

20th and 21st centuries have seen unprecedented growth, with the highest growth rate occurring in 1963 at a peak of 2.2 percent per year.[204] By 2011, the global growth rate had declined to 1.1 percent.[205] In 1900, 1.6 billion people inhabited the Earth. By the year 2000, that number had increased to more than six billion.[206] At present the Earth holds over 7.034 billion people.[207]

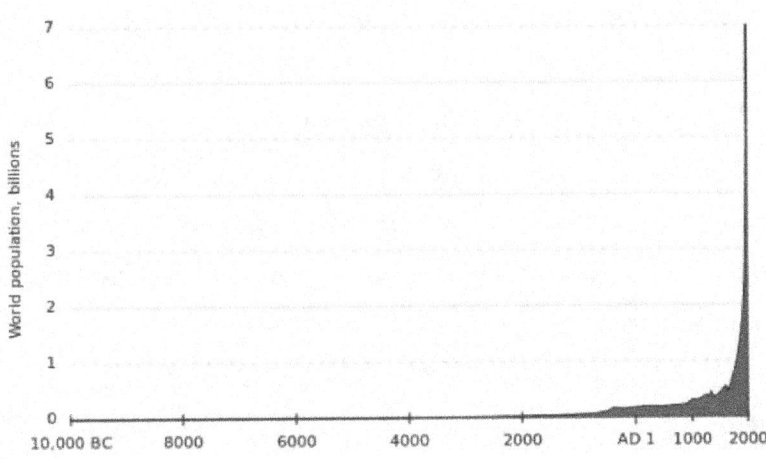

Figure 12.5 The shape of human population growth, from 10,000 B.C. to 2000 A.D.[208]

Over the course of human history, countless emergent patterns of human social organization, from political hierarchies to global trade routes, can be traced to patterns of energy flow along underlying attractor pathways, i.e., oscillatory (Hamiltonian) paths of least resistance for the flow of energy dissipation. From this perspective, human societies are complex adaptive systems that have evolved to facilitate the extraction, consumption, and dissipation of energy through the human superorganism. Intensive agricultural practices have deepened these channels over time, redirecting much of the energy that flows through Earth's ecosystems into the information and energy exchange networks that constitute human societies. The evolution of complexity in human societies is the progression of this energy flow, organized in autocatalytic sets of socially ordered activity.

12.9 The First Agricultural Emergence

Like all human beings, the people of the Levant, a region of the eastern Mediterranean, lived in foraging societies organized at the family level. Agriculture emerged from these circumstances as gardening supplemented foraging activities, and plants and animals were gradually "domesticated" in the sociocultural system over time.[209] Living in a region of extreme climatic shifts between cold rainy winters and hot dry summers, the Natufian society invested surplus energy toward the construction of semi-permanent settlements.[210]

The social network structure of the Levant evolved in close association with the patterns of relative energy abundance. Carriers of myths and legends performed ceremonial gift-giving rituals (distribution of surplus energy) and funeral rites (dissipation of energy). Specialist elders stored and transmitted knowledge orally, acting as important nodes in the social information network. An elder woman (45 years old), identified by archeologists as a shaman, was discovered to have been buried with ritual items including the pelvis of a leopard, the wing tip of a golden eagle, the tail of a cow, marten skulls, the forearm of a wild boar arranged parallel to her own forearm, and the shells of over fifty tortoises (these last consumed during a burial feast).[211] [212]

Shrines were constructed, designating certain public areas as ritual spaces.[213] The most famous structure from the period is the tower of Jericho, located in one of the world's first towns, Jericho (in modern Palestine), which then had a population of 2,000 to 3,000 people.[214] The stone tower and its mud wall had communal ritual purposes. The solstice sunset cast the shadow of a hill to the West onto the tower, connecting it with the cycling of the seasons. It may also have served to protect against invaders and flooding.[215] The tower was an early example of an emergent monument, the architectural manifestation of surplus energy channeled through a society producing beyond subsistence needs in a bottom-up, communal manner.[216]

New social hierarchies and energy surpluses fed back into early Mesopotamian society, positively reinforcing technologies that increased carrying capacity. As sedentary populations cultivated local grains, including barley and wild oats, in concentrations greater than could be disposed of by way of consumption and ritual sacrifice, the excess energy was gathered in granaries controlled by households or individuals.

The first such surplus energy storage system appeared around 9500-8500 B.C., in Jericho.[217] [218] Granary construction was a cooperative venture, constellating existing family structures into networks organized around shared resources and the production and storage of surplus capital.[219] Under these conditions, permanent settlements and villages of roughly three to five hundred individuals grew along the shorelines, cultivating annual plants.[220]

12.10 Ancient Global Exchange Networks

The evolution of human societies proceeds in accordance with the deepening of energy distribution channels over time. Complexity arises as a function of the efficient integration of information and energy exchange networks that emerge through the distribution and consumption of energy. As societies organize in autocatalytic sets of socially organized activity, enduring structures emerge that propagate the human superorganism through time and space. To illustrate this process, we will now examine the emergence and evolution of trade networks, which have evolved over the course of millennia, from footpaths following the path of least resistance between ancient settlements, to flight paths following geodesic arcs between modern cities.

In ancient times, amber was a valuable raw material that served religious ritual functions in various societies. It has been used in jewelry since as long ago as 11,000 B.C.[221] Bits of amber accompanied the Egyptian pharaoh Tutankhamun in his burial chamber, having been brought to him from mines near the Baltic Sea. The Amber Road originated in multiple locations, near the locations of amber mines.[222] Initially, connecting roads were nothing more than informal pathways taken by individual traders searching for the path of least resistance between one concentrated energy source and the next. According to thermodynamic law, energy flowed from regions of high energy density to regions of low energy density. In this context, human beings acted as the carriers of Gibbs free energy through the system. As traders exchanged information about the best routes through trading circles, more people traveled and cleared the roadways, creating a positive feedback loop of entrained activity that formed the Amber Road. Just as the accumulation of water, coursing through a watershed, leads to the emergence of a river, the Amber Road emerged through the prolonged activity of human travelers seeking paths of least resistance. This maximized the dissipation of concentrated energy sources by exchanging goods and services throughout the socio-energetic network.

The organizational pattern we find in trade routes was not designed in a top-down manner. As people continued to use the routes, or energy and information distribution channels, the networks of activity self-organized and reinforced their structures, regardless of the specifics of the primary commodity being traded. By 3000 B.C., the Amber Road became the Tin

293

Road, as the Bronze Age tin exchange network simply transplanted itself onto routes previously established by Amber Road traders.[223] [224] This network subsequently merged with another, when between 2000 and 1500 B.C., traders from Central Asia joined the tin exchange, expanding the network along an east-west axis by way of the Silk Road, and from there, along existing lapis lazuli trade routes.[225] This dynamic network of interconnecting energy and information exchange continues uninterrupted into the present.

Every major roadway first emerged under conditions determined by the circulation of energy and information in the form of goods and services, only later to give way to another system of energy exchange, as determined by the relative scarcity and abundance of food sources, as well as other types of energy and their symbolic equivalence as rare metals. Modern trade routes are but the latest iteration of the perpetual flow of information, people, and energy through the ages. The evolution of complex trade networks, alliances between societies, and even empires proceeds through this process of exaptation: the co-option of an existing phenomenon to perform a new function.[226] Biologists use this term to describe adaptive forms preserved by natural selection that are subsequently used for a different function. Here we extend its use to describe the evolutionary development of the human superorganism.

12.11 The Royal Road and an Example of Emergent Empire

During the Achaemenid period around the 5th century B.C., an energy and information network emerged through the interactions of roughly 50 million people, centered in present day Iran.[227] For roughly two centuries, this vast Persian Empire extended east-west from the Aegean Coast to the Indus Valley, and north-south from the Iaxartes/Syr Darya River in the north to Aswan.[228] The society was composed of a vast trade network of energy and information exchange at a scale previously unknown.

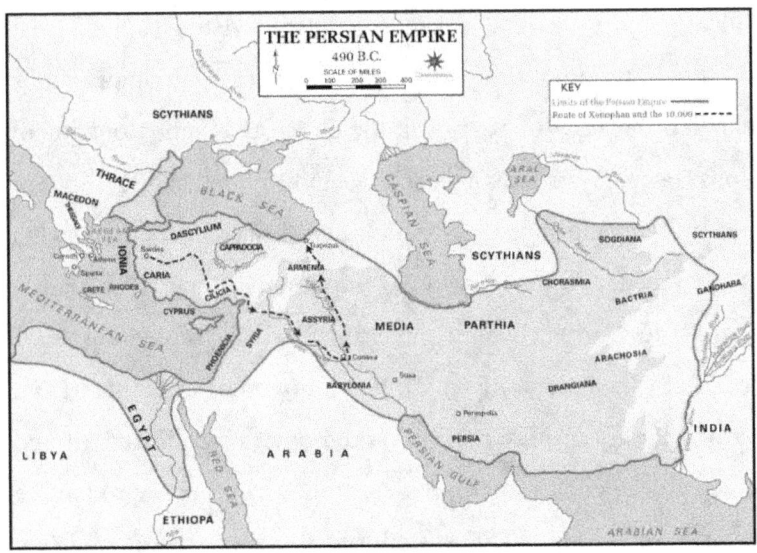

Figure 12.6 Territory of the Achaemenid Empire at its greatest extent, 490 B.C.[229]

Elam's position as a central node in the network of goods and services extending into neighboring territories, with

Mesopotamia to the west and the Indus Valley to the east, made it the trading crossroads of the "cradle of civilization."[230] Elam was a highly connected node in the "global" trade network of the period. The coherence of commerce throughout such a vast network emerged through the positive feedback channels of an autocatalytic network composed of multiple urban centers united by a shared system of organized religion and bureaucratic institutions. The flow of energy through this network became possible through the increased efficiency of information exchange, which was itself made possible by cuneiform writing.

The empire emerged after several states were integrated politically by a war that subsequently carved the path of least resistance for the flow of energy. These attractor paths channeled the flow of energy in the superorganism. Constituent nations became increasingly connected by exchange brought about by war and marriage, which constellated tribute systems that provided surplus energy to support the social structure. Under King Darius, twenty government nodes called satrapies were established, each of which had a governor appointed from the royal family or high nobility. The governor collected a fixed tribute from each satrapy, determined by the productive capacities of the land. This system of tribute extraction, codified on cuneiform tablets, redistributed surplus energy and wealth throughout the networked nodes of the empire.

The king traveled the land, collecting tribute organized by the individual territories. For a few months each year, the sovereign would occupy one of the royal lodges that had been constructed along the Royal Road, oversee the collection of tribute, and then travel to the next royal lodge, according to a seasonal cycle. Each satrapy maintained warehouses along the roads called thesauroi: food depots set aside for the royal caravan. Ancient inscriptions on the Persepolis Fortification tablets recorded the supply transfers from individual satraps to the royal caravan: "Sixty-four marris [about 170 gallons] (of) first class clarified butter, supplied by Mastetinna, were consumed before the king, at Susa and five villages (humanus), in the twenty-second year."[231] The tablets, which documented precisely where travelers stopped on the road for food, also implied a complex system of credit and payments extending throughout the empire.[232] [233] Royal travel brought with it the flow of specialized information, as experts including masons, goldsmiths, and herders accompanied the king as he circulated throughout the network of interconnected road systems.[234]

During this period of imperial emergence, a rigid etiquette system emerged, which forbade people from standing in the presence of the peripatetic king, who mandated that they prostrate themselves before him.[235] Rulers often recited their royal lineages, thereby acting as channels for concentrating and reflecting cultural information. Information flowed from the people to the rulers in a bottom-up fashion. For example, Cyrus, a predecessor of Darius, did not impose religion in a top-down fashion. Rather, populations practiced their own local religions, and when priests at an Apollo shrine in Asia Minor gave a prophecy favorable to Cyrus, he granted the priests special privileges—an example of a positive feedback loop.[236] [237] When Cyrus died, his son, Cambyses, began making regular offerings to Cyrus' soul at his tomb, including the daily sacrifice of a sheep.[238] According to Herodotus, Zoroastrian sacrifices could be made only in the presence of an information specialist called a magus.[239] The attractor pathway governing information and energy flow was redirected from Cyrus to Cambyses.

When Achaemenid society interpreted the works of Darius, it did so as a bottom-up expression of the superorganism's constituent individuals, spanning various cultures. Darius' "autobiography," which recounted the arcane information of his genealogy, was inscribed on a rock face at Bisotun in three languages: Old Persian, Elamite, and Babylonian.[240] Darius widely republished records of rebellions, their causes, and the causes of his successes, feeding information back into the society that selected him as its emergent leader.[241] A new writing system, created by a group of Persian scholars for imperial ceremonial purposes only, reflected the new, imperial level of complex hierarchy. Inscribed monuments and a tomb with religious carvings recorded the process by which the culture's religious symbols (flame, winged circle, royal seal, sun, Akkadian moon-symbol, disk with crescent) and political symbols (carvings of the six noble Persians who served Darius) converged on the leader, who became a reflecting lens for the entire society. Darius' tomb was not the only emergent expression of the culture's religious and political information. The tombs of all subsequent Achaemenid rulers were adorned with the same symbols with slight modifications. This process of enshrinement is analogous to the sacrifices of Cambyses to his father, Cyrus, and the Hawaiian chiefs' lineage recitations. The principal difference between the Hawaiian and Achaemenid systems is scale, and the medium of information transfer, oral culture versus a culture with higher population, writing, and enough material wealth for the emergence of large-scale monumental architecture.

The remarkable similarities in social structure, modes of energy and information exchange, and the development of institutional hierarchies reflected in religious symbols, practices, and symbolic architecture associated with an elite class, are testament to the evolutionary principles underlying the emergence of complexity. Each region of agricultural origin was contingent upon the particular characteristics of the local environment, and in regions of abundant energy, complex agricultural societies emerged in the wake of the agricultural revolution. The diversity of expression exhibited by these societies, reminds us of the diversity of species Darwin observed in the Galapagos, with each particular society an expression of its unique evolutionary history contingent upon these regional variations. That such broad similarities have been observed in the development of social complexity, speaks to a fundamental underlying mechanism: the selection of cultural expressions as the most energy efficient forms able to utilize channels of existing energy flow. The attractor pathways of energy flow that characterize complex adaptive systems are self-similar across scales, and are exhibited by networks of biomolecules, organisms, and human beings. The most energy efficient forms are preserved along their passage through channels carved by thermodynamic flows of energy, captured in information networks, and preserved in the formation of complex adaptive systems.

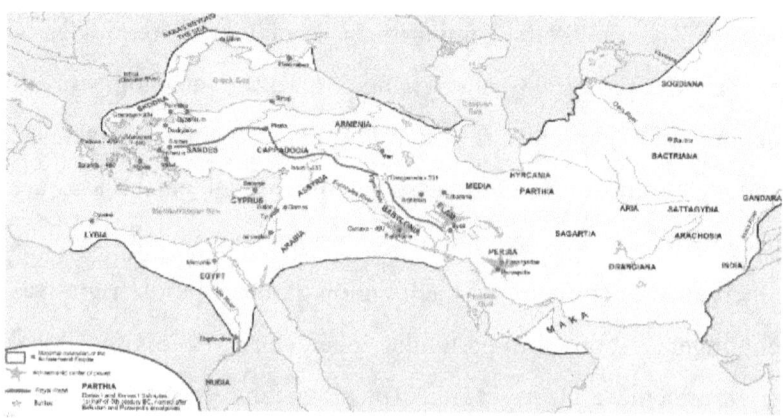

Figure 12.7 The Royal Road during the Achaemenid Empire, 5th century B.C.[242]

12.12 Language and the Flow of Information from East to West

Following the dissolution of the Western Roman Empire, the population of Europe throughout the Early Middle Ages (400-1000 A.D.) was relatively stable, between 25 and 30 million, repeatedly checked by ecological limits and military clashes.[243] Before the translation of Arabic sciences into Latin during the 12th century, populations in Europe had grown according to rates determined by ecological conditions, as well as the limits of subsistence agriculture. By the 14th and early 15th centuries, knowledge of Arabic in Europe resided mostly with a few medical doctors interested in the source material for their practices, and with Italian merchants who traded with the Levant. Here we find that not only goods and services flow along trade routes. Information does as well. Arabic took the path of least resistance into Europe.

The first center of translation was built in Toledo, Spain, during the 12th century. Here, John of Seville and others translated the ancient works of mathematics, astronomy, astrology, philosophy, and medicine from Arabic, assisted by a native Arabic speaker know as a *Ghalib* (Galippus).[244] Information from a text translated from the Tasrif—a 13th century Arabic book by the surgeon Al-Quasim Al-Zahrawi— became an important source of medical knowledge throughout European universities during the High Middle Ages.[245] Along with specific techniques of chemical distillation, crystallization, and the use of alcohol as an antiseptic, westerners acquired the basis for the scientific method from these texts. Though much systematic knowledge, primarily the texts of Aristotle, had been transmitted from the Hellenistic world, empirical techniques that later characterized the scientific method were transmitted through the human superorganism by Arabic translators. Perhaps the most important contribution of Arabic thinkers was that of mathematics: number theory, algebra, and the concept of zero were all transmitted to European society by way of Arabic translators. Technologies essential to seafaring, which would become critical to the development of the modern world, were circulated throughout European universities during this period. The energy and information exchange networks enabled by these technologies further deepened the attractor pathways that contributed to the Industrial Revolution and the emergence of the "Modern West." Seafaring inventions included the lateen sail, mariner's compass, and nautical charts. Finally, technologies related to print-printing, including the manufacturing of paper, spread into the West through Arabic traders by way of Spain and Sicily, creating a positive feedback loop for the distribution and invention of knowledge itself.[246]

Advanced techniques, including the scientific method, spread from Arabic into Latin, rather than the other way around. By sharing a common language, the Arabic world was much more integrated, and information flowed more easily in the Arabic medium. Before the conquest of Spain in the early 8th century, Islamic armies had spread throughout the Arabian Peninsula, into Syria, Egypt, and Iran. As those territories became economically integrated, energy and information flowed along the paths of least resistance—identified by common cultural symbols and systems of communication. Houses of translation emerged as nodes in a network of information and energy exchange that converted the knowledge created by various civilizations into Arabic. In Baghdad, in the year 830 A.D., the Bait-ul-Hikma, or "House of Wisdom" was constructed, where Muslim scholars translated philosophical and scientific works from Greek, Syriac, Sanskrit, and Pahlavi (the rarified, scholarly language of pre-Islamic Iran) into accessible, widely spoken Arabic. Knowledge came into the Arabic network over several centuries, from the Assyrians, Chaldeans, Babylonians, Phoenicians, Egyptians, Indians, Iranians, and Greeks. This bottom-up process of translation and piecemeal acquisition turned Arabic into the world's most extensively disseminated scientific language.[247]

The Western scientific method, it appears, had far from rational origins. Key components of the method emerged through decentralized energy and information exchange channels developed in the wake of Muslim invasions. Hardy and starving hordes of Arab horsemen flowed out of the arid Arabian Peninsula in search of new resources, attracted by the energy abundance of the more sedentary societies of southern Europe. A number of positive feedback loops fed from the culture at large back into the processes of knowledge acquisition that occurred at the Bait-ul-Hikma. For example, Muhammad encouraged Muslims to learn to read and write, and literate captives of battle could win their freedom by teaching, reading, and writing to ten children from Medina.[248]

New information generated at the Bait-ul-Hikma positively reinforced the culture as well. Muslims used astronomical observations and calculations to determine times for prayer and fasting, as well as to properly orient themselves toward the Qiblah, in the direction of the Kaaba in Mecca to pray. Mathematics and geometry were implemented to determine tithes and tribute. "Revelation writers of the Quran" began copying books. The culture rewarded those who could copy accurately and completely. Information, itself a form of energy, was an enabler for the transfer of more energy. Literacy increased as more texts were translated into Arabic, beginning with the scientific texts that dealt most closely with people's daily activities. Libraries, universities, and houses of wisdom were constructed alongside mosques, and staffed by people from different religious and cultural backgrounds. These new centers were supported by the state treasury.[249] These large-scale patterns of organization embodied by Muslims, not only united the Muslim superorganism under the law of God, but also laid the energy-information infrastructure for many other societies of varying degrees of complexity. This specialization is reminiscent of the properties that emerge by way of the coordinated cellular activities of a multicellular organism: eating, digestion, and finally the circulation of energy in the central nervous system, which in turn feeds information back in order to better capture food energy and avoid danger. Here, at the level of human societies, we see the coordinated activity of an entire population linked by bonds of energy and information exchange again leading to the emergence of a new complex adaptive system.

12.13 Maritime Economies and the Emergence of Industrial Europe

Perhaps more than any other factor, the westward flow of information and energy catalyzed by Arabic translators during their Renaissance of the 12th century was responsible for the emergence of imperial European powers during the Early Modern period. It is not surprising that Spain emerged as the first such powerful nation. Translation, and the diffusion of technology through the Iberian Peninsula, situated Spain as a powerful Mediterranean commercial hub at the center of the West. By the 16th century, following the Crusades, significant maritime innovation, and the expansion of trade routes between the East and West, gradually shifted from the Islamic Mediterranean toward the North Atlantic and the increasingly integrated Atlantic maritime network.

Just as cuneiform writing on clay tablets provided the information content that enabled the vast expansion of trade networks in the Ancient Persian Empire, so too did the printing press and printed broadsheets enable the emergence of robust trade networks throughout Europe at the dawn of the Modern era. In the 16th century, Dutch merchants began printing business newspapers that allowed for the circulation of information about prices and exchange rates, previously available by word of mouth alone, to anyone able to buy a newspaper. Enabled by print technology, the Antwerp Exchange emerged as a central hub of economic information and activity, integrating the economic networks that provided the energy and information basis for the emerging European empires. Newspapers announced lending rates, the arrival of cargoes, and the departure of ships. By the 17th century, the Commodity Price Current was published in at least four languages: English, French, Italian, and Dutch. In 1688, the Dutchman William of Orange became William III, King of England. He brought with him a court of Dutch merchants and financiers to London. In his wake, techniques of information consolidation and dispersal spread to Britain, and the commercial hub of the West shifted from Antwerp to London. By 1716, a London merchant could subscribe to seven different financial weeklies. Likewise, there was huge demand in the colonies for pricing information from London.

By the 18th century, London had become a highly urbanized node in an expansive, networked global economy. Transportation technologies such as shipping permitted the flow of goods (energy), information, and people throughout the world. The steam engine had been invented in Britain, as part of the emergence of manufacturing and what historians would later call the Industrial Revolution. By the 18th century, expanding literacy enabled a broader base of the population to participate in the manufacturing economy. To support this economy, people and energy clustered into urban centers, moving from rural environments and agricultural professions into the cities. The centuries between 1500 and 1800 saw the emergence of the most expansive empire the world has ever seen, united politically under the banner of the United Kingdom, and constituted by a number of networks of information and energy exchange that grew increasingly interconnected during the age of exploration.[250] By the early 18th century, one quarter of Londoners worked in shipping or related port services.[251] Populations further congregated in the commercial center of London, as technology-driven agricultural productivity and urbanization positively reinforced one another in a powerful feedback loop that has defined human growth trajectories ever since.[252]

The Industrial Revolution emerged as print and transportation technologies rapidly diminished the energy expenditure required to consolidate flows of information, energy, and individuals into increasingly integrated trade networks. As those networks expanded through the same processes outlined in our discussion on the emergence of the Amber Road, certain nodes benefited greatly from the network effect. The trading empires of the Dutch, Portuguese, and British, enabled by print and transportation technologies, built global network connections with far-flung colonies on every continent.

Networks of information and energy exchange that would later become part of the British Empire had been expanding long before the emergence of British maritime dominance. To illustrate, let us consider maritime Asia between the years 1500 and 1800. Asia was itself highly integrated by 1500, with trade networks connecting the Indian Ocean, the eastern Mediterranean, and Persia. Since 900 A.D., China had exported large quantities of porcelain and silks throughout Eurasia by way of a network of shipping routes that extended throughout Southeast Asia and the Indian Ocean. Indians exported silks and fine cotton to Arab territories and East Africa. Within India, "portfolio capitalists" invested in handicraft, commercial, and tax-farming initiatives that integrated trade networks across multiple regional courts.[253] When Portuguese traders traveled to India in the wake of Vasco de Gama's voyage to Calicut in 1498, they followed pre-established Muslim trade routes extending from Mozambique to Southeast Asia. Indian rulers often granted revenue collection rights to certain Portuguese merchants and societies, including the Society of Jesus. The result established an integrated Indian-Portuguese network.[254] Similarly, linkages between the British East India Company and Bengal multiplied during the 18th century. As the Mughal Empire disintegrated, Indian nobles hired French and British soldiers to train their armies, granting them special revenue privileges. British traders expanded their activities from India into China, where they sold guns, opium, and tea to all of

Southeast Asia.[255] By 1757, the East India Company had emerged as Bengal's preeminent economic power.[256]

Far from being an isolated phenomenon of imperial expansion ex nihilo, the 18th century British Empire was an emergent expression of a global trade network that merged long-established Asian trade routes with the rest of the world. London became the hub of innovation for this network, developing efficient practices of information exchange that facilitated both the integration and expansion of existing trade networks. England's privileged role in this integration was a function of its adaptive culture, which produced specialists of integration. Bills of exchange facilitated exchanges of products between Londoners and merchants in the West Indies, Iberia, and Ireland. Local networks of trust emerged that facilitated exchange among specific groups throughout the Empire. Americans and Britons crossed the Atlantic to orchestrate industrial production processes. Servants of the Dutch East India Company traveled to India to gather information from Indian artisans through emissaries known as *banians* and *dubashs* ("men of two tongues"), who acted as interpreters and cultural interlocutors.[257] London merchants operating in the Atlantic formed entire supply chains, increasing efficiency of investment in American plantations, the slave trade, and military contracting. The cumulative effects of millions of repeated individual actions, entrained in networks of energy and information exchange through commerce, together led to the emergence of the British Empire.[258]

12.14 The Island Effect

Just as the environmental constraints of a circumscribed island environment contributed to the emergence of the complex Hawaiian State from its origins as a Polynesian chiefdom, so too did the island environment of England contribute to its emergence as a powerful industrial empire. The increased population pressure of an urbanized people in search of new energy resources impelled the British to venture beyond the European continent. The population density of cities such as London and Manchester increased the rate of innovation in military, industrial, and transportation technology, leading to inventions such as the cannon, the steam engine, and the modern naval vessel. This internal population pressure, coupled with the abundance of Gibbs free energy provided by the island's coal deposits, provided the expansive forces necessary to propel a society of seafaring adventurers and merchants into a prominent position in an emerging global information and energy exchange network.

Internal pressures within the limited geography of the English isles provided an ideal environment for rapid innovation, as well as for the selection of robust naval vessels.[259] The fully rigged ocean ship, developed and refined by European shipbuilders from the 15th century onward, was for centuries the

310

cheapest and most-efficient form of transporting goods over long distances. Its design was continually refined, until steam technology made it obsolete.[260] Once on the open seas, British traders carried their cultural and trade practices over long distances, unimpeded by local populations. Long before Europeans began to dominate Asian territorial trade in the 1750s, European ships had already established control over long-range Asian maritime trade routes.[261] In the New World, European ships conferred enormous military advantages upon invading Europeans, enabling armies to congregate in larger numbers than could any local, slower moving native population.[262] At the same time, shipments of colonists acted as an escape valve for growing European populations.

The limited geography of the small island nation put an upper limit on its inherent carrying capacity as well as on its expanding population. As Britain's population grew, war acted on the population in both negative and positive feedback loops. The Napoleonic Wars at the turn of the 19th century solidified Britain's grip on maritime power. As Napoleonic France expanded its influence over the European continent, and the United States emerged as an independent nation in North America, Britain focused on the seas. As Britons left the confines of the island for the high seas, population pressure was reduced in British cities, where a surplus in labor had emerged.[263] During the Napoleonic Wars, London became a safe haven for skilled workers and capital from mainland Europe, perhaps because of the difficulty of attacking Britain across the English Channel.[264] Isolated from the chaos on the European continent, England became a hotbed of the development of several industries that contributed to both the efficiency of agricultural production, and the expansion of industrial technologies.[265] Technologies developed for warfare were applied to the agricultural sector, helping to usher in a "second" agricultural revolution that accompanied the urbanization of British society.[266] Armed with these innovations, agriculturalists expanded into waste, scrub, and unenclosed land, enlarging the society's energy base in concert with the transformation of its internal structure. Coal production doubled to meet wartime energy needs, and after the war this increased productive capacity became available to the domestic sector of the economy. Finally, British naval dominance of the world's oceans united global trade under a single set of laws, greatly facilitating the efficiency of information and energy exchange. In retrospect, the consolidation of national identities around their relative positioning in global trade networks, industrial and agricultural production, and systems of government unmoored from specific geographic locations, constituted a major shift in the self-organization of the human superorganism: the advent of the modern state. As the site of the major innovations in trade relations, industrial production, and energy economy, Britain was a uniquely configured microstate that served as the model for the macrostate reorganization of human societies at the global level during what became known as the *Pax Britannica*.[267]

12.15 Shift in Energy Production

We have repeatedly observed how populations grow to the carrying capacity of their environments, determined by an abundance of Gibbs free energy available in forms to which the society has adapted. Throughout history, social innovation has repeatedly occurred in cultures forced to adapt efficient energy and information exchange networks in "islands" of energy scarcity. Major innovations in "social metabolism," and the corresponding changes in social structures that emerge to deal with energy scarcity, radiate outward from their island origins, as populations equipped with these innovations venture out from isolated environments. Throughout history, these innovations have engendered islands of energy scarcity, whether in continental arid regions such as Arabia and Mongolia, or actual islands such as Britain and Japan. In search of additional sources of energy, the innovations developed in relative isolation radiate outward from these societies. In a sense, the innovations in energy and information technologies that emerge from these regions function as attractor paths around which greater macrostate changes constellate. We have repeatedly seen how critical innovations, whether adopted by Mongolian and Arabian horsemen, or Polynesian and British mariners, became cornerstone features of the energy and information architecture

313

of larger complex societies. Just as the Polynesian agricultural and maritime toolkit served as a cultural attractor for the Hawaiian Chiefdom, Arabic translations and scholarship provided the materials that led to the European Renaissance. In much the same vein, the British industrial and mercantile culture became the model for the modern state.

Perhaps the most significant British innovation, in terms of its long-term consequences, was the emergence of an economy dependent upon the combustion of fossil fuels. When the "metabolic" activity of British society reached the energetic carrying capacity of the British Isles, a radical transformation in the society's energy base occurred. Between the 16th and 18th centuries, in parallel with the expansion of its mercantile and seafaring activities, as well as with the clearing of its forests for grazing, British society adopted the use of coal as a domestic source of fuel. By the beginning of the 18th century, roughly 50 percent of British coal was consumed in household fireplaces. By the end of the century, nearly all thermal energy was derived from coal. Individual homeowners and inventors modified the design of coal-burning homes piece by piece, adjusting individual elements such as grate material (metal or brick), fireplace size, fireplace material, chimney taper, chimney flutes, ventilation system, etc. Because these were not the sorts of innovations that could easily be patented, technological improvements in energy production circulated readily in urban London. This "collective innovation" on the eve of the Industrial

314

Revolution was evolutionary and occurred through bottom-up innovation rather than top down organization and planning. In the face of energy scarcity, society reorganized itself by adopting innovations to consume a novel energy source, found locally in abundance and unknown elsewhere in the world.

An agricultural revolution emerged alongside this early phase of the British Industrial Revolution, mutually reinforcing the growth of cities. This revolution, too, was an outgrowth of new patterns of information flow. The rapidly diminishing cost of printed materials, in a maturing, literate society, fostered an enormous increase in agricultural innovation, as measured by the number of publications of agricultural patents and technical farming books.[268] Farming books were extremely effective vehicles for the spread of information and increased literacy among the population, which facilitated the transmission of the new techniques.[269] Thus, Britain was ideally situated to emerge as the locus for the radical increase in invention, as well as the adoption of industrial technologies.

In the wake of the Industrial Revolution, societies adopted patterns of development that continue uninterrupted to the present. Interestingly, the infrastructure of social metabolic networks—whether of the agro-economic structures of the Hawaiian chiefdoms, the satrapies of the Persian Empire, or the megalopolises of modern states—expands sub-linearly with respect to population density. In other words, as human societies become more complex, they become more energy efficient per population unit.

Just as Kleiber's allometric scaling law applies to the metabolic rate of individual organisms, it also applies to the energy and transportation infrastructure of human societies.[270] Bettencourt and others found that the energy storage capacity, road surface area, and length of electrical cables used for urban infrastructure, scale according to the power law that characterizes the allometric scaling of basal metabolic rate in biological organisms. As societies become more complex, they become more energy efficient. In other words, the complexity of a society's economy is selected on the basis of its energy efficiency.[271]

In the context of the urbanization of human societies, however, something unexpected emerges. Bettencourt and his team discovered that creativity and innovation—measured by number of patents issued, research and development budgets, and 'super-creative' professions, among other factors—scaled in a manner that also indicated a quarter-power law. However, the slope of the scaling was super linear (greater than 1) with respect to energy efficiency. A city ten times larger than another is not just ten times more innovative; it is 17 times more so. A metropolis 50 times the size of a small town is 130 times more innovative.[272] As discussed in previous chapters, Kleiber's law reflects that as living organisms grow larger and more complex, metabolism per unit of body mass slows down. Bettencourt's research measured a crucial principle of scaling in human societies: as human populations grow denser, they generate ideas at an exponentially faster rate. On average, a resident of a metropolis with a population of five million people is almost three times more creative than the average resident of a town of a hundred thousand. This super linear increase in innovation and creativity is a concrete example of the information network effect on the energy efficiency of the human superorganism. It may also be correlated with the lower metabolic rate of larger and more complex organisms, whose increasingly interconnected nervous systems provide a "network effect" of innovation that leads to increased energy efficiency in other components of the metabolic system.

12.16 Thermoinfocomplexity in Human Societies

The fractal patterns of complex adaptive systems extend from the molecular level into the realm of human society. Human societies are adaptive, dynamic systems that select for the most energy efficient social structures in the face of energy scarcity at a variety of levels of energy density and corresponding complexity. At the microstate, family foraging behaviors follow seasonally fluctuating activities that self-organize along attractor pathways of energy distribution, as we observed in the !Kung and other foraging societies. These patterns of activity scale upward through all levels of social complexity. Initially, patterns of energy exchange self-organize

317

within the reciprocal bonding of kinship networks, which become more complex in the face of population pressure. A shift from egalitarianism to hierarchy occurs as population pressure leads to the intensification of resource use as societies expand. Though the total amount of Gibbs free energy used by a growing society necessarily increases as the population grows, the efficiency with which it transforms that energy into the architecture of its social network, increases as the society becomes more complex. In the face of energy scarcity, a society bifurcates into hierarchies that enable the formation of larger social networks, as it adapts to new sources of Gibbs free energy. These transformations become structurally integrated through the many forms of "ritual attractors" that emerge in complex societies. Although diverse in form and content, these "ritual attractors" provide the basis for social bonding beyond the purely biological relationships of family, kin, and tribe, and are essential to the information architecture that undergirds complex chiefdoms, states, and empires.

From prehistory to the present, networks exchanging energy, information, and individuals carve out attractor basins, through which societies evolve. Societies become more complex as they encompass new energy, and information flows through exchange networks that bind human beings together via the dynamic interplay between the forces of competition and cooperation. The flow of energy through these attractor basins has increased over time as these networks bind more energy into social structures capable of sustaining greater human population densities. The massive flow of information from the Arabic world at the end of the Late Middle Ages, the integration of global maritime networks by the European colonial powers during the Early Modern period, and the information and agricultural revolutions of the Industrial era have continually evolved as the human superorganism grows in size. In the current information technology revolution, all of these subsystems have now merged into an integrated energy and information network that is truly global in scope.

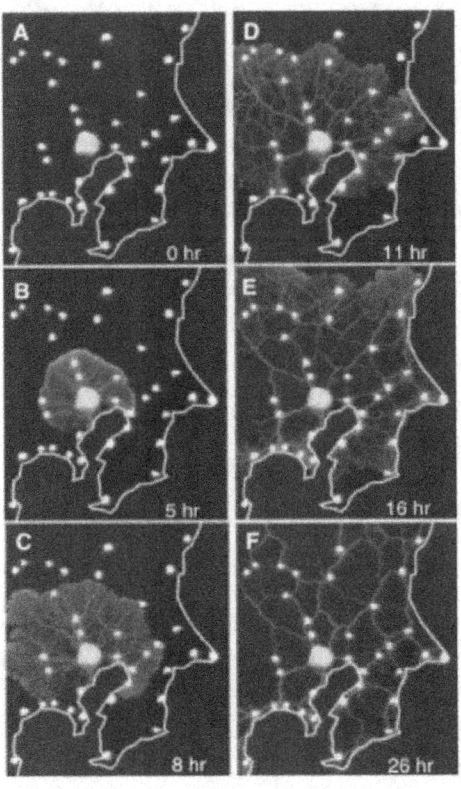

Fig. 1. Network formation in *Physarum polycephalum*. (**A**) At *t* = 0, a small plasmodium of *Physarum* was placed at the location of Tokyo in an experimental arena bounded by the Pacific coastline (white border) and supplemented with additional food sources at each of the major cities in the region (white dots). The horizontal width of each panel is 17 cm. (**B to F**) The plasmodium grew out from the initial food source with a contiguous margin and progressively colonized each of the food sources. Behind the growing margin, the spreading mycelium resolved into a network of tubes interconnecting the food sources.

Figure 12.8 Slime mold path mimics Tokyo Area Commuter

319

Rail.[273]

Despite the remarkable diversity of human societies, the organizational patterns of human populations are remarkably self-similar across all scales of complexity. Modern-day Japan is perhaps a paradigmatic contemporary example of a complex modern society, continually evolving on an island of relative energy scarcity. Japanese researchers, working with the familiar slime mold, have demonstrated in a remarkably elegant manner the coherence of the scale-free evolutionary patterns that emerge in complex adaptive systems. Working with a "map" of Japan, which included topological barriers that mimic the geographic terrain, the researchers placed food sources for the slime mold in the same relative locations as Tokyo and 36 of its neighboring cities. The resulting network of the slime mold's extended plasmodium produced a remarkable image of the commuter rail system used by the human superorganism that is Japanese society.[274] In a way that is consonant with Thermoinfocomplexity theory, Tero and colleagues argue that just as the slime mold is able to determine the shortest path through a maze by seeking the optimal, metabolically efficient network, so too do human societies, when viewed as an emergent superorganism. The slime mold builds highly efficient networks in a self-organized manner, through the elegant process of energy and information exchange by means of positive and negative feedback loops in the extended plasmodium.

At first glance, it may appear that the patterns of organization that we observed at each level of complexity may be directed in a top-down manner—an organism by its brain, or a society by its leader. In reality, the brains of organisms and the leaders of a society are nodes, albeit important ones, in a complex network of energy and information exchange, operating throughout the complex adaptive system. Given the remarkable patterns of self-organization and optimization that we have traced throughout the evolution of complex adaptive systems, from molecular networks and foraging societies to the most technically and industrially advanced societies, we can only imagine that the contemporary emergence of a globally interconnected computational and information communication network presages yet another shift in the emergent complexity of the ever-evolving human superorganism. In the next chapter we will use the principle of Thermoinfocomplexity theory to predict and imagine the trajectory of human social organization as it evolves in a coevolutionary path with its broader ecosystem the biosphere of planet Earth.

Chapter 13 – Imagine

13.1 The Trajectory: Human-Computer Information Network System

Predictions about the future of human civilization rely on understanding the interactions of an enormous number of variables. However, armed with the knowledge of how stochastic interactions of energy, information, and matter lead to the emergence of increasingly complex adaptive systems throughout time, we can visualize the complexity of life and human society through the novel lens of Thermoinfocomplexity theory. In the previous chapters, we observed emergent patterns throughout nature, from molecules to human societies. We have learned that emergence is unpredictable in all its details, yet the future is an outgrowth of the past and the present. We are poised to make plausible predictions about the future of human social organization; of course, we are aware that this path is not deterministic and it has an inherent level of uncertainty.

How may complexity theories provide insight into what the future will hold? There was an article featured in the December 2011 issue of *Scientific American* by David Weinberger, a senior researcher at Harvard University's Berkman Center for Internet and Society, which raised the question that this chapter addresses in more detail. Other complexity scholars are now suggesting a variety of approaches. Dirk Helbing, physicist and chair of sociology at the Swiss Federal Institute of Technology at Zurich, proposed a plan for a Living Earth Simulator, which would take into account data from every area of human activity—economic, sociological, agricultural, techno-logical—and then make suggestions for policymakers based on its calculations. The European Commission (the executive branch of the European Union) ranked Helbing's plan first, among six finalists in a competition to receive one billion euros. At the Center for Complex Networks Research at Northern University, Director Albert-László Barabási and colleagues have developed a model that predicts with 90 percent accuracy, where a person will be at 5 p.m. tomorrow, based on past behavior alone. This is a far cry from predicting the future of global human society. Nonetheless, it engages the same issues as Helbing, exploring the regularities in human systems to predict the future.[275]

Although these efforts are immensely laudable, they all share a major blind spot. They are attempts at designing a very complex system, to advise leaders using a top-down approach. As we have described in some detail throughout this book, the evolution of complex adaptive systems in nature is quite the opposite. Evolution is a bottom-up process with no designer. It is spontaneous, elegant, and surprising. Therefore, the project of gathering extensive data to predict one's likely location or personal preferences, in order to advise leaders for the top-down design of an efficient socioeconomic system, is not really based on an understanding of how adaptive complexity emerges and progresses. The precise numbers of interacting variables and their future sequence in the complex global network of human beings, in the context of other species and the changing physical environment, is beyond predictability. No top-down system could harness all the trillions of probable variables in a stochastic, dynamic process of energy and matter interaction adaptable to the ever-changing biosphere. The biosphere is not amenable to a predictable top-down design. The surprising emergence of complex adaptive systems can only be explained by the principles of Thermoinfocomplexity, which are based on the interactions of energy and information. These stochastic interactions lead to the emergence of complex adaptive systems, which include the global human social organization.

Therefore, in the following pages we look at the future of global society through the lens of bottom-up emergences. We predict with great uncertainty and humility that there will be a new complex adaptive system of globally connected human beings through extensive communication networks. This prediction, although logical, is only one possible outcome. We will refer to this complex emerging system of unprecedented energy efficiency, as the interactive computerized algorithmic network (ICAN). We define ICAN as: *An interconnected information and energy network that includes humans, computers, and machines. This network will be selected for its energy efficiency following the shortest route (Hamiltonian). The network of human-social organization emerges from the bottom-up progression of the complex, self-organized, adaptive, and robust society.* This system is in the process of forming from a patchwork of independently emerging computerized information networks. More and more, these networks are merging to improve efficiencies in the production and distribution of goods and services throughout the world. In addition, the networks are becoming increasingly interactive with humans through positive and negative feedback loops. This is creating a hybrid human-computer structure. The artificial intelligence itself is an emergent complexity that will merge with the vast network of human brains, resulting in a much more extensive and efficient human social organization. Of course, this is all built on the foundation of the evolutionarily derived, complex human biological structure that includes the human brain/consciousness. The resulting complexity of artificial intelligence and human consciousness will be an exciting prospect for the emerging human global society (Gaia).

Consider that at least 99 percent of the computational capacity of personal computers connected to the internet is unused at any given time.[276] The merging of multi-level AI may be considered a consequence of this computational power. To start, let us first examine the most important contributor to this emerging network: the human brain. How did humans develop complex inventions that no other animal has been able to achieve? About 10,000 years ago, the rise of specialization among individuals and the extensive communication between them led to the emergence of agricultural society, which provided the foundation for increased productivity and rapid population growth. In turn, the increased communication in heavily populated communities geometrically increased the rate of innovation in those communities.[277] This led to the industrial revolution of the late-18th to early-20th centuries. We are now at the beginning of an information and communication revolution.

13.2 The Brain as the Seat of Intelligence

The human brain is the most complex information processing system in the biosphere. It forms the foundation of our emergent consciousness and the intricate information exchange facility within our bodies, governing all we think, feel, and do. It is also essential to our sociality and communication, and it is the main contributor to our evolutionary success. As early as the 4th century B.C., the Greek physician Hippocrates (perhaps best known today for the physician's Hippocratic Oath, "Do no harm") believed the brain to be the "seat of reason." But the brain wasn't always understood this way. The ancient Egyptians considered the brain to be nothing more than "cranial stuffing," to be removed promptly during mummification; they instead considered the heart to be the seat of intelligence. Aristotle felt similarly, believing that the heart was the seat of intelligence, while the brain was merely an organ for cooling the blood (paradoxically). Aristotle concluded that the exceptionally large human brain, which rapidly cooled down hot-blooded impulses, was responsible for the high degree of rationality of man compared with beasts.[278]

Today we know that human intelligence springs from the complexity, size, and organization of the brain. Hippocrates was right: the brain turned out to be the seat of reason. Humans have more cells within the cerebral cortex (11.5 billion cells) than any other animal, with the exception of elephants and certain whales. By comparison, a mouse has about 8 to 14 million cortical neurons, a cat about 763 million, and a chimpanzee 8 billion. The cortex is responsible for much of the information processing in the human brain. It is involved in the association of inputs and the higher reasoning ability we find in humans, a feature that is apparent in less-complex forms in chimpanzees. So, what makes our brains so smart?

13.3 The Foundation of Intelligence

Intelligence, which we define as the behavioral and mental flexibility with which an organism adapts in response to a changing environment, is not simply a reflection of the number of neurons or the overall mass of the cortex. After all, elephants, consistent with their much greater size, have a much higher brain mass than human beings, and possess roughly the same number of cortical neurons, if not more, as is the case with many of the larger cetaceans.

So, what accounts for this variability? The organization of the brain is as important a contributor to the information processing power of the entire neural network, as is the size of the brain and the number of neurons firing within it. In humans

328

in particular, several factors make the human brain the most powerful biological information-processing machine. The human cortex is thicker and contains a much higher density of neurons than that of any other animal. These neurons are the interacting agents that make up the web of communication and exchange responsible for the emergence of consciousness, as well as our higher cognitive and social functions. The neurons of the human brain are also highly myelinated, which means they are covered in a sheet of fat that surrounds an individual nerve fiber and greatly enhances the conduction velocity of a signal traveling along the axon of a firing nerve. This enhances the speed of communication within the neural network. In humans, cortical fibers are thickly myelinated, enabling rapid information exchange, whereas the myelin sheaths of elephants and cetaceans are thinner and less effective. Moreover, the human brain has a tremendous number of synapses, or tiny gaps between neurons where information exchange occurs. Each neuron in the human cortex has about 29,800 synapses, resulting in about 360,000,000,000,000 cortical synapses. That's 360 trillion synapses, each of which, in turn, relies on a chemical brew of neurotransmitters to communicate electrical messages across the synaptic gap. The higher conductive velocity, the smaller distances between neurons, the large number of synaptic and dendritic connections, and a large variety of neurotransmitters (about 60 or more) give the human brain a substantial edge in information-processing over all other living things and account for our higher intelligence.[279]

In short, the human brain is in essence a carbon-based adaptive computer with a high degree of plasticity. This plasticity is embedded in the atomic structure of the carbon atoms. The carbon atom is unique in that it has four electrons in its outermost orbital, so it is able to bond to a variety of atoms easily (for example carbon-oxygen bonds). But it can also break those bonds easily to bond with another element or compound. It is the promiscuity of the carbon atom that has put it as the foundation of emergence of all life. The brain learns, adapts, and changes its network architecture in response to its physicochemical environment. In essence, the plasticity of the human brain makes it different from a super-fast silicon-based supercomputer, no matter how many trillion computations the computer might be capable of doing. The silicon-based computer is not adaptive to changes in its environment. Adaptation is essential to what makes the human brain so unique and powerful.

13.4 Complexity and the Human Brain

Over the last 40,000 years, the human species has come to dominate every corner of the earth through an explosion in specialization and information exchange. In the previous two million years, our species underwent an exponential growth in brain size. The graph below indicates the expansion of cranial capacity among hominids. 1.8 million years ago, the cranial capacity of members of the *Homo* genus was 702 cc (an average of skulls from Perning, Kenya, and Dmanisi). Since then, cranial capacity has increased by six standards of deviation.[280]

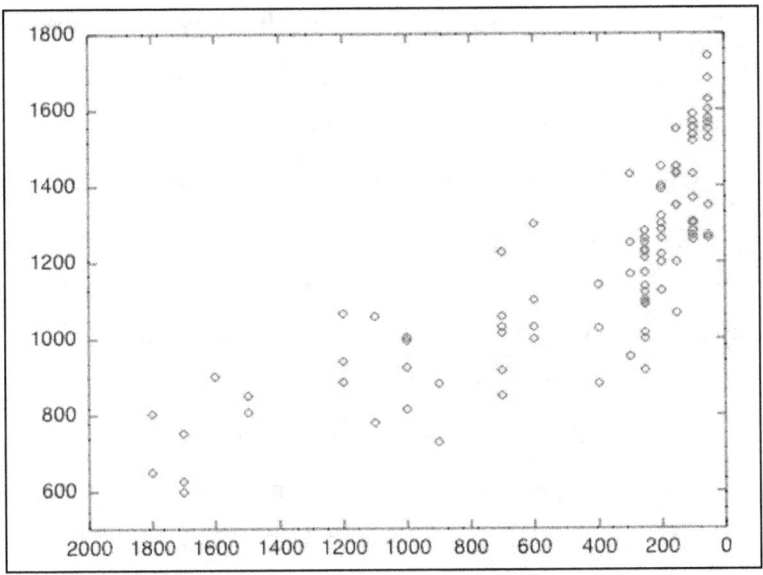

Figure 13.1 Cranial Capacity (cubic centimeters) vs. Time (thousands of years ago).[281]

The tissue of the brain is the most metabolically expensive part of the human body, representing only two percent of the body in terms of mass, but requiring roughly 25 percent of the body's glucose utilization, 15 percent of its cardiac output, and 20 percent of total oxygen consumption.[282] As energy efficiency is a filter of the evolutionary selection process, brain size is dependent upon the amount of available

Gibbs free energy in relation to the advantage of its intelligence (information processing power).

Over the last 20,000 years, the human brain has reversed its steady growth in volume so that now the average size of the male human brain is 1350 cc.[283] And the female brain has about the same volume. This shrinkage is roughly equivalent to losing a golf ball sized chunk of brain matter. This has been more than compensated by the efficiency of packing neurons into a smaller volume, which has made their connections closer, more rapid, and more energy efficient. Some postulate that over the last 10,000 years the individual human brain has shrunk due to disuse, compensated by cooperating individuals in the agricultural society. I think that this is too short a time for evolution of smaller brains. As I have said above, the reverse is true. A more compact and smaller human brain has resulted in more efficient communication within it. In fact, migration toward better sources of food resulting in the agricultural society is a consequence of more intelligent individuals searching for better sources of food.

There is no evidence that larger brains necessarily correlate with greater intelligence. In fact, the brain of Albert Einstein was a third of a pound lighter than the average male brain. Studies of Einstein's preserved brain indicate that there may be structural features allowing for increased communication and, hence, greater intelligence.[284] The lateral sulcus, a fissure found in all brains, is much smaller in

332

Einstein's brain than in average brains. It has been speculated that neurons in this area could have communicated more easily and rapidly. As it turns out, this area of the brain is known to be the site of mathematical and spatial reasoning; and indeed, Einstein was a visual thinker. Though size does not necessarily correlate with intelligence, complexity does. Perhaps the complex nature of Einstein's highly specialized brain, contributed to his scientific breakthroughs. By contrast, the largest brain ever recorded weighed 4.51 lbs. and belonged to an idiot: a person with an IQ below 30.[285]

Recall that theoretical chemist John Avery established that information and energy can be correlated as a ratio of one electron volt of energy per every 56.157 bits of information. The implication of efficient communication, within individual human brains and between them in a tightly knit structure, is profound. Whoever can convert energy into information most efficiently, and effectively communicate this information, will be more likely to survive. So, although the size of the human brain has decreased over the last 20,000 years, this does not imply that our intelligence has decreased. In fact, it may have increased. Or, like the brain of ants, the human brain may have become more specialized and energy efficient, allowing for a lower BMR/g among individuals, in addition to greater processing power in specific temporal regions. As the population density of a society increases, the specialization of individuals leads to the more-efficient use of energy throughout the entire network. Moreover, within the shrinking brain itself, dendritic connections become more densely clustered and hence, more efficient in transmitting information. It may appear as if Moore's law, a pattern governing the increasing processing power of computers, has been at work within our own brains, albeit at a much slower rate.

13.5 A Happy World with Minimal Government

In the chapter on ant colonies, we saw a very complex superorganism in the form of collectives working, evolving, and persisting over hundreds of years in a very efficient manner, without any perceivable bureaucracy or government providing guidance or distributing resources among individual ants. But compared with humans, ants had a distinct time advantage in the development of their leaderless society. It took over 110 million years for ants to develop a complex system of seamlessly coordinated, individual specialization. Through super-specialization embedded in their genes, each specialized ant communicates with others through chemical cues (pheromones), contributing to social organization.[286] None of this is conscious. Though each ant has only about 250,000 neurons, collectively, the colony functions with a much higher degree of intelligence.

Other similar examples of collective communication and super-specialization can be found within our own bodies. Our various specialized cells communicate with each other in multiple ways. Individual cells live out their normal lifespans, like individual ants and, through their specialization, contribute to the much longer average lifespans of human organisms. All cells communicate with each other through chemical and electrical signaling and feedback loops to maintain the homeostasis of the entire body. When comparing human societies to ant colonies, we observe that our stochastic process of evolution has taken us down an adventurously more complex path. That path was made possible by the large number of cells in our bodies. Specifically, the colony of our body holds about 10 trillion cells and another ten times that many bacterial guest workers, plus 60 quadrillion symbiotic mitochondria.[287 288 289 290]

To maintain energy efficiency in such a large menagerie, extensive and rapid communication between the individual, specialized members emerged in the form of an extensive communication network.[291] Our complex brain consists of roughly 100 billion cells devoted solely to communication within our body which is, in essence, a multicellular colony. In the evolution of complex adaptive systems, complexity plateaus, and then as it evolves further, often jumping to a new, emergent state that is qualitatively different from the previous one.

As a society, our species has had only roughly 50,000 years to evolve into a significant social organism. In fact, the evolution of cities and larger social communities is only about 10,000 years old, originating with the dawn of agriculture. In the last 10,000 years, through the expansion of language, writing, the printing press, and now communication through the internet, the game has changed substantially. We are now at the threshold of making a qualitative jump to the next level. We are witnessing the beginnings of the emergence of a heretofore unpredicted social and economic structure.

I predict that even if we experience the calamitous destruction of a large number of people through war, disease, and starvation, another serendipitous "happy ending" may be awaiting the human species. That happy ending will be manifested through a collective and efficient information-exchange network that is now gathering force and rapidly becoming possible among individual agents in our global society. This process is witnessed in the advances in computer technology, rapid visual and auditory global signaling, and the application of complicated algorithmic solutions to major economic and social problems. It should go without saying that the rapid growth of the internet is just the beginning of this transformation. To give us a glimpse into the possibilities of our future, the next few sections will briefly describe a number of independently emerging, computerized information networks that each independently and automatically solve major problems with greater-energy efficiency. The internet will function as the evolving, four-dimensional scaffolding that will support many independently emerging solutions to society's problems. This information network, which includes at least seven billion interacting human brains, will coordinate all the interactions of our human species. Eventually, the human species will function seamlessly, just as an ant colony or the cells of our own bodies operate and communicate in a bottom-up manner. This collective communication network has now manifested itself in the growing network of artificial intelligence, at this time silicon-based, interacting with over seven billion carbon-based computers: our individual human brains. Our global information network will accomplish all this in a much more complex and aesthetically elegant manner. The human brain, with all its complexity, is an important and central node in the emerging, interconnected information network of our global society. Following are a few concrete examples of emerging nodes throughout various communication networks.

13.6 Computerized, Algorithmic Solutions for Consumer Trade: Walmart

In 1962, the Walmart Corporation began as a humble supermarket founded by Sam Walton in Bentonville, Arkansas. What made this company different and successful was the realization that major cost (energy) efficiency was possible through a "just-in-time" computerized inventory and ordering system. Whenever an item is purchased from the Walmart checkout aisle, the computer automatically notifies the warehouses and the distributor for its replacement. In time, Walmart's new inventory control system led to a computerized, networked supply chain. The purchase of a single item sends information to each node in the chain. In the 1980s, economists used to claim there was an insurmountable three day barrier between selling and restocking—but not anymore.[292] Walmart no longer has to send someone down the aisles to mark the items in need of restocking on a clipboard. When an item left the store at the checkout counter, the supply chain was automatically instructed to replace it.

Another way that Walmart now uses bottom-up, networked technology to improve its efficiency is its automated scheduling process. In the past, Walmart employees would work standard shifts, full-time or half-time. But ideally, the number of workers in a store should reflect the number of customers demanding service at a given moment. Before networked information systems were available, managers had to guess how many employees should be present for the morning or the afternoon shift. By examining patterns of actual customer traffic through the checkout point, Walmart automated the scheduling of its workers in the most efficient manner. Workers submit their availability and an algorithm sorts out which employees come in when, in order to meet real demand for customer service. Computerized work assignment increased efficiency by eliminating a number of middle managers. Walmart's automated inventory management has been so successful that the software, developed by Kronos, has been used by other companies as well, including Ikea, Fossil, and New Look.[293]

The philosophy of using computers to weed out inefficiencies has spread throughout Walmart's entire corporate structure. In response to a growing reputation for environmental irresponsibility, Walmart announced in 2005 broad reaching plans to increase its energy efficiency. It was an attempt to answer criticism that the environment was suffering from the externalities of Walmart's business practices. Here, we note that another element, namely reputation, worked as a feedback loop to influence the behavior of the corporation, which ultimately led to energy efficiencies (and more profit) for the company. We will come back to reputation and its importance soon; suffice it to say that while Walmart fulfilled the needs of its customers, the impulse to try a new, energy efficient strategy began with negative feedback issued by people aware of Walmart's inefficient, environmentally destructive practices. So, Walmart took action. By eliminating excessive packaging on a line of kids' toys, Walmart saved $2.4 million in shipping costs. Furthermore, it created its own electric company to power its stores. In the process, the company saved 3,800 trees, one million barrels of oil, and about $15 million per year. It also improved its reputation among its customers. Further, by simplifying its trucking lines, Walmart laid down plans to reduce its greenhouse gas emissions by 20 percent in seven years, saving money and reducing environmental damage.[294] [295] [296]

Now that we've discussed a few of Walmart's networked approaches, let us take a look at the energy and monetary advantages to said approaches and examine how they mimic the superorganisms we find in nature. For one, a Walmart store can stock a lower inventory and still meet customer demand. A networked restocking system weeds out the inefficiencies of guesswork present in the traditional approach. This traditional approach favored smaller stores with lower overhead costs, such as lighting and cleaning, and a store with more diverse products able to sell goods to a diverse customer base. Walmart's networked system allows it to take greater advantage of fads and other uncertain-to-sell products. For example, in the past, retailers such as J.C. Penny would order clothing 90 days ahead of when it would be displayed in the store. No one knew if the clothes would still be in fashion when they hit the racks. This old-style of retail was so inefficient that entire industries sprang up to capitalize on the goods that fell through the cracks. T.J. Maxx, Ross, and Loehmann's represent "bargain basement" stores, which help move the retailer's older, unsold goods. Such discounters will become increasingly obsolete in an age when all retailers can respond to what their customers actually want at any given moment.

Speaking of cost, it is well known that one of the largest expenses incurred by any corporation are the wages of its employees. By computerizing its inventory and automating its scheduling system, Walmart has been able to secure greater value from its labor costs. By combining efficiencies with bulk purchases, Walmart has been able to pass on savings to its customers.[297] A 2007 study showed that Walmart's price reductions saved consumers $287 billion in 2006. Customer savings amount to $957 per person, or $2,500 per household. In 2009, Walmart paid its workers a combined total of $933.6 million in bonuses, and an additional $788.8 million in profit sharing and 401k contributions.[298]

As the computerized information network has expanded, middle management has shrunk. Walmart is able to do away with a large number of employees whose jobs have been computerized. In its delivery of goods and services, Walmart is more efficient per remaining employee. This is called "disintermediation" (or "getting rid of the middle man"). We can see that in the history of Walmart, the growth rate has increased continuously, whereas in most corporations, as the sales of the corporation increases, the number of employees also increases and the corporation gradually reaches static profitability until, eventually, its profitability decreases.

The accumulation of wealth (surplus energy) at the top of the hierarchy is an interesting phenomenon observed in the structure of all human social hierarchies, from the !Kung in Africa to the monarchs of Hawaiian society to the super-rich of the modern era. We may speculate as to the function or dysfunction of this type of energy (money) accumulation in a small part of the social superorganism. Observations indicate that at some point, the extreme concentration of energy in a small part of the organism, without redistribution throughout its body, will result in the demise or substantial reorganization of its structure.

It is evident that when networked computers match supply and demand perfectly, or use algorithms to calculate resource allocation, the added efficiencies are incredible and contribute to the customer's well-being. All this is well and good but, at the moment, the redistribution of profits is dependent on the top-down decisions of the corporate leaders of Walmart. There is currently no system in place within Walmart to ensure that resources are distributed to maximize the well-being of workers, customers, and other stakeholders. It is true that some money has been passed onto the workers and some to its customers, but a disproportionately large amount is still flowing to the wealthy elite within Walmart's management, shareholders, and other institutional investors. In 2010, Walmart's net sales totaled $405 billion, with a net profit of $24 billion.[299]

The five most prominent members of the Walton family, who together own 39 percent of the company, are worth about $18 billion apiece, for a total of $89.5 billion.[300] In 2010, the CEO of Walmart, with a salary of $30 million, made almost more in one hour than an entry-level Walmart employee earned in a year. The CEO's hourly wage came to $16,826.92, compared with the $8.75 per hour for new employees, totaling an annual entrance-level salary of $18,200 based on a 40-hour work week.[301] This skewed distribution of wealth, most likely does not contribute enough to the well-being and happiness of those few superrich to justify its continued implementation. In fact, it may be viewed as a source of inefficiency, which so far, none of the present networked systems have addressed. We may find that as Walmart looks for ways to increase its efficiency and profit (using bottom-up technology whenever possible), other more-efficient, information-based algorithmic solutions may emerge that can lead to a more-efficient distribution of its resources, to the benefit of all involved.

13.7 Computerized, Algorithmic Solutions for Health Care: Dr. Watson

In health care, we are witnessing the emergence of another algorithmic information network. Like Walmart, the health care industry is evolving into a radically new, more efficient system, as it begins to see new forms of networked information put to use in patient care. In 2011, Watson, a computerized network of knowledge, defeated the all-time highest Jeopardy! money winner, Ken Jennings, as well as the contestant with the longest-ever winning streak, Brad Rutter. Watson had access to 200 million pages of information, totaling four terabytes of disk space.[302] Watson was able to process 500 gigabytes of data every second—the equivalent of one million books.[303] Using massive parallel processors, Watson ran a variety of algorithms based on the words and syntax of the trivia questions and then assessed the certainty of its answer. In its programming, the more algorithms that independently developed the same answer, the more likely Watson would deem that answer to be correct.[304] Despite wordplay, ambiguities in syntax, and expert opponents, Watson handily defeated its human contestants. Watson won the game with $77,147, a sizable victory over Jennings' $24,000 and Rutter's $21,600.[305] In defeat, Ken Jennings made a foreboding remark: "Just as factory jobs were eliminated in the 20th century by new assembly-line robots, Brad and I were the first knowledge-industry workers put out of work by the new generation of 'thinking' machines. 'Quiz show contestant' may be the first job made redundant by Watson, but I'm sure it won't be the last."[306]

Since that televised game of Jeopardy!, IBM has applied the technology behind Watson to the field of health care. Dr. Eliot Siegel, professor and vice chairman of the Department of Diagnostic Radiology at the University of Maryland's School of Medicine, noted that "as a physician or radiologist, it might take me 10 or 20 or 60 minutes or more just to understand what's in a patient's medical record. Watson can ingest information efficiently and rapidly. It'll have an encyclopedic knowledge and suggest diagnostic and therapeutic possibilities based on databases much larger than one physician can possibly hold in his head."[307] Watson could be a terrific tool for physicians, summarizing a patient's health record, or pointing out potential problems stemming from adverse reactions to drugs. This diagnostic and therapeutic tool will work hand-in-hand with the growing, networked databases of health care information. Another radiologist, Dr. Arun Krishnaraj, commented that "We must be honest with ourselves. Currently, computers augment our imaging capabilities, but we know that one day computers will be faster and more accurate at making diagnoses than we are."[308] One writer went so far as to give Watson the catchphrase: "The Second Opinion You Plug into a Wall."[309] The more that information is recorded, networked, and made available to algorithms capable of bringing the relevant bits together, the more efficient our health care system will be.

In health care, efficiency saves not just money, but also lives. According to an extensive meta-study conducted by Gary Null, et al., as many as 797,000 Americans die every year as a result of poor medical treatment.[310] Shockingly, if this figure were listed as a cause of death by the Center for Disease Control, preventable medical errors would be the leading cause of death in America.[311] Those deaths also entail an economic loss of over $239 billion.[312] But, what if electronic health records (EHR) were completely networked and accessible by point-of care diagnostic software like Watson? With automated diagnosis and streamlined access to patients' EHRs, hospitals will not have to perform as many medical tests, especially highly invasive exploratory surgeries, which are partly responsible for America's medical tragedies. Adjustments in medication and new diagnoses will be arrived at by "Dr. Watson" and relayed to the treating physician or directly to the patient. A network of sensors interacting through a mobile device, which in turn interacts with a server and provides alerts to patients and medical professionals, would no doubt save money and lives. Imagine if the 24 million Americans with diabetes had instantaneous access to their glucose and insulin levels, streaming in real time. On a large scale, such a system would offer enormous benefits to society at large. A massive, interlocking public health dataset would emerge, revealing previously undiscovered trends and synthesizing feedback from patients. Soon, the algorithmic interactive network may combine with drug research and discovery, as well as consumer safety concerns throughout the supply chain.

13.8 Computerized, Algorithmic Solutions for Intelligent Transportation Systems and Social Networks

In the realm of transportation, we are witnessing the emergence of network intelligence. Intelligent transportation systems (ITS) are beginning to implement the same techniques as Walmart and the medical sector to improve the efficiency of roads throughout the world. Like Walmart, ITS use algorithms and actual usage data to increase efficiency, making moment-by-moment adjustments to conditions on the road. ITS have begun to take advantage of newly available volumes of data. The drive for ITS began with the problem of how to move cars in the most-efficient manner possible. In 2011, American drivers in urban areas traveled an extra 4.8 billion hours because of traffic congestion.[313] Sitting in traffic, they burned an extra 1.9 billion gallons of fuel at a cost of $101 billion.[314] In the U.S., the average rush-hour commuter spends one full week in traffic each year.[315] ITS would largely eliminate these inefficiencies, allowing for a transportation revolution.

Millions of cellphones distributed throughout the population provide data for this transportation-information network. Even when turned off, all cell phones periodically transmit their locations to their mobile phone networks. The technology behind the cell phone network was not invented with traffic management in mind, but this increasingly ubiquitous communication device helps a traffic management system track the movement and location of cars that determine the flow of traffic. By analyzing these emergent patterns via algorithms, the network is able to make precise self-improvements. Roadways are now monitored by networked technologies, including GPS devices, license-plate readers, geographic information systems (GIS), fiber-optic networks, wireless networks, and variable speed limit systems, among other instruments.[316] In Japan, sensors in the road automatically prompt electric signs to indicate that a car is stalled on the highway. Networked data may even measure the frequency with which drivers tap their brakes. As a result, we may render a stunningly accurate, real-time depiction of traffic congestion.

As data continues to grow in a bottom-up fashion, transportation planners have the opportunity to automate

347

management and take full advantage of the new network paradigm. The old way of controlling roads relied on the same mindset as that of a Walmart manager deciding in advance that x number of employees work in the morning and y in the afternoon. But, what if predictions on the number of drivers were even more dynamic? On Britain's M25 motorway, electronic signs with variable speed limits auto-adjust to road conditions, including congestion and rain, by demarcating lower and upper speed limits. The results, since 1995, have shortened commute times, reduced accidents, and produced smoother traffic patterns. The U.K. highway agency estimates that traffic congestion costs the country £3 billion each year.[317]

In the U.S. and elsewhere, cameras have been installed that automatically ticket vehicles exceeding the speed limit or disobeying red lights. Such automatic ticketing establishes an immediate feedback loop and has been shown to reduce accidents locally, by an average of 40 to 50 percent.[318] One study showed that by applying real-time data to the timing of traffic signals in the U.S., and automating the timing according to the changing data, stops can be reduced by up to 40 percent, total travel time by 25 percent, and gas consumption by 10 percent. These efficiencies would save 1.1 million gallons of gasoline every year, cutting carbon emissions from vehicles by 22 percent. One networked system in Tucson, Arizona, required an investment of $72 million to set up, but showed a return of about $455 million from increased efficiency—a return on investment ratio of 6.3 to 1. Applied to the entire United States, similar estimates suggest a ratio closer to 25 to 1. In other words, a $1.2 billion investment across the U.S. would see a $30.2 billion return.[319] Moreover, these gains would be realized as savings to individual drivers, rather than to any one person in charge of monitoring the network.

In Minneapolis, Minnesota, simply metering an onramp reduced the number of crashes at the site by between 15 and 50 percent. In 2005, ITS reduced delays in U.S. urban areas by 9 percent, giving drivers back a whopping 336 million hours, representing $5.6 billion in increased productivity and fuel. In Singapore, an automated traffic control system has already reduced traffic congestion by 13 percent, and increased average vehicle speed by 22 percent.[320] Although the U.S. has been somewhat slow to adopt ITS relative to other developed countries, these transportation systems are projected to reduce the number of American traffic fatalities, injuries, and associated costs by 50 percent.[321] France and Germany have already heavily implemented ITS to great effect. In 2006, the European Union committed $212 million to implementing its eSafety program, which connects vehicles, signs, and drivers in a massive network consisting of advanced vehicle sensors, GPS, and real-time electronic maps. In France, an extensive toll-road infrastructure, known as Liber-t, is now in place, collecting driver data at a central server.[322]

Data traversing multiple nodes of the interactive algorithmic network have many applications. With data gathered by an increasingly networked and intelligent transportation system, the government can better direct an evacuation in the event of a foreign threat or natural disaster. Homeland Security employs operators at ITS control centers to monitor "critical infrastructure," such as bridges and tunnels, for suspicious activity. The domains of automated traffic management and security are beginning to merge, giving rise to reduced operational friction and increased efficiency for both systems.[323] As society grows increasingly networked, the efficiency of the entire emergent system will grow exponentially.

Finally, a vast array of social networking sites is another node in the growing information network, bringing social and business relationships together in an overlapping system. Games and economic apps proliferate on social networking sites, bridging what began as a platform for person to person interaction within the realm of business, and even politics. Like each node in the emerging network, Facebook and Twitter increase the efficiency of those who use it. With the emergence of democratic revolutions throughout the Middle East, we witnessed the greater ease with which collective action may be affected via social media. These movements represent a compelling example of individual human brains coordinating massive action through a silicon-based computer network. The proliferation of a diverse array of social networking websites has allowed the instant dissemination of information between people of different social classes in various areas of the world. These sites inform individuals about changes in news or other events, in some cases, before they can be reported by a top-down news media outlet. In a few instances, people have posted messages about an earthquake before it hit their friends in a neighboring city. This combined computerized algorithmic network will increase the adaptability and efficiency of the global human superorganism. We are now rapidly developing our global brain through AI and its multi-level development.

13.9 A Value-Free, Bottom-Up, Evolutionary Future System

So far, we have mentioned four examples of automated, computerized algorithmic solutions for increasing efficiencies in the areas of consumer trade, health care, transportation systems, and social networks. We have also shown that interconnection and information exchange between these nodes of the imagined network of global communication are mutually beneficial. As the interconnection grows, so does the efficiency of all the connected systems, following the general rules of network effects. We have discussed the above examples in some detail, in order to illustrate how interactive computerized algorithmic networks (ICAN) can increase economic and energy efficiency for large numbers of human beings, freeing their time for other pursuits. Numerous other areas may contribute to increasingly efficient global communication, including education, and the collaborative exchange of creative activities in the arts, science, technology, food production, music, gaming, entertainment, and so on. These networks and more are evolving as we speak. Through cross connections they will continue to drive the productivity of the entire system.

We have intentionally postponed our discussion of perhaps the most interesting, humanistic, and evolutionary aspect of self-organizing, complex networks. This discussion predicts the inevitable emergence of complex, collective negative and positive feedback loops, stemming from the huge number of individual humans interacting with the system. In our discussion of Walmart, we described how feedback networks are made responsive to continuous input at the checkout counter. The system we will now describe represents a new emergence, as the nodes of human social networks become ever more interconnected. This future system is free from human judgment and an a priori set of values. It is a bottom-up emergent evolutionary process in progress. The system we will describe is simply based on a long trajectory of the emergence of complex adaptive systems, progressing toward energy efficiency following a Hamiltonian process.

Imagine if the algorithmic system, as a whole, were subject to constant and continuous input by individual humans for all decisions and outputs. We believe that in the future, through interactive human inputs, an all-encompassing mechanism will instruct and inform the continuously changing and emerging information network, communicating its instructions to the various segments of the social superorganism. This network will be informed by and responsive to the wishes and desires of all humans on a 24/7 basis. It is a real-time, true democracy, if we need to call it that. Any change would reverberate throughout the whole network, evoking further adaptive responses. In that sense, this future society will be directed and controlled by humans through billions upon billions of "useful" and "not useful" input signals. Its actions will be both individualized and communal, informed by the decisions made by all people on earth. Clearly, such an interactive information network will serve the survival and stability of the human species which, I might add, requires the preservation of the Earth's crust, atmosphere, environment, and its intricate biodiversity. A unified ICAN network would govern human society and affairs through an efficient, distributed, bottom-up communication system that works without leaders. With every emergence in evolution comes a cost at the transition points. There may be social turmoil and suffering for some, as well as wars and other challenges, as humanity becomes a fully networked society.

13.10 Artificial Intelligence and Human Society

This vision of our future is in stark contrast to the predictions of Ray Kurzweil, I. J. Good, Vernor Vinge, and similar futurists. Kurzweil predicts that "the singularity is near," the title of his 2005 book predicting the future of technology. He defines the singularity as the point in history when human intelligence is outstripped by artificial intelligence. "The singularity," Kurzweil writes, "does not achieve infinite levels of computation, memory, or any other measurable attribute. But it certainly achieves vast levels of all of those qualities, including intelligence."[324]

According to Kurzweil, computers will achieve the complexity of the human brain through some combination of molecular computing, the self-assembly of nano-scale circuits, DNA based computing, and perhaps quantum computing, which employs atomic "circuits" operating at the smallest and most powerful levels known to modern science. Vinge describes the process in the following manner: "[w]e will soon create intelligences greater than our own. When this happens, human history will have reached a kind of singularity, an intellectual transition as impenetrable as the knotted space-time at the center of a black hole, and the world will pass far beyond our understanding."[325]

Kurzweil elaborates on the prediction. He states that "[t]he singularity will allow us to transcend these limitations of our biological bodies and brains...There will be no distinction, post-singularity, between human and machine."[326] Does Kurzweil mean that silicon-based robots will come to be self-programming, reproducing computers? Will they take their own path, making us a sideshow that can be discontinued and cast

354

aside as early as the year 2045? Perhaps not. These speculative projections are just that—speculative hypotheses that do not meet Karl Popper's requirement of falsifiability, which would be required to make the scientific theories worth pursuing.

By contrast, the future I imagine has a very rational, evidence-based foundation. It is the result of observing what is actually happening, and what has already happened throughout the process of evolution, from social superorganisms, like colonies of ants and bees, to multicellular organisms, including our own human bodies. Many futurists, including Kurzweil, believe that advances in artificial intelligence will continue to occur at an exponential rate. It is true that breakthroughs in computer technology continue to follow one another over shorter and shorter intervals, as per Moore's Law. However, I doubt that any silicon-based computer technology will have the plasticity and stochastic flexibility to surpass the time-proven, carbon-based biological information network. Carbon-based networks continue to evolve through molecular complementarity and intricate communication. The human carbon-based brain is unlikely to be surpassed individually, and even less likely as a unit in a growing network of more than seven billion humans across the globe integrated into an interactive, silicon-based global computer network. The power of the emerging networked system, ICAN, comes from the interconnection of these computer networks emerging across disparate fields.

Unlike Kurzweil's prediction, the ICAN model does not rely on the expansion of individual human intelligence via machines. ICAN gets its power from the bottom-up, across all sectors of human activity, and its feedback loops operate on a global scale. The single best adaptive and plastic carbon-based information network in the ICAN system is the human brain. Yes, carbon-based AI may replace silicon-based AI, but it will integrate with the human global brain that created it in the first place. I predict that biology will always drive technology. It's the interaction of many brains within the ICAN system that is so powerful. It is not any particularly enhanced node within the network that drives the network efficiency. Such a carbon-based, evolving neurochemical signaling network, interacting with the carbon-based interactive computerized algorithmic network (ICAN) would be able to set the course, not only for human interactions, but for any sort of robot activity imagined by the futurists.

13.11 Carbon versus Silicon

Now, let us ask the obvious question: why didn't silicon evolve into a life form? The answer is simple—it is not nearly as versatile an element as carbon is. In an atom, only the valence electrons can form chemical bonds. Carbon and silicon both have four valence electrons, and can form up to four bonds. Because these elements are right next to one another on the periodic table, they contain the same number of electrons in their outer orbital. Hence, one might expect many of their bonding properties to be the same.

Yet, silicon lacks the ability to form chemical bonds with as many different types of atoms, limiting the chemical versatility needed to establish a working metabolism and thus the emergence of self-regulating, complex adaptive systems. Large carbon-based molecules often contain hydrogen, oxygen, phosphorus, and sulfur, but large silicon molecules are often quite monotonous compared to the great variety of organic molecules.

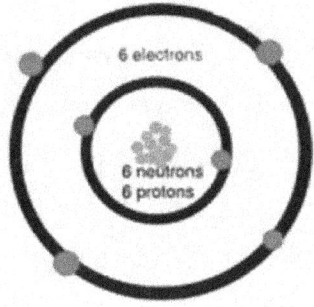

Figure 13.2 Orbitals of a carbon atom

If silicon and carbon have such similar electron configurations, why are the molecules that they form so different? The main difference is size. Carbon's nucleus is made up of six protons and, usually, six neutrons. Silicon, on the other hand, has fourteen protons, and an average of fourteen neutrons—over twice the amount, which means over twice the mass. While carbon only possesses six electrons (four of which are valence), silicon has fourteen. The ten electrons closer to the nucleus change silicon's chemical properties and cause its atomic radius—the radius from the middle of the nucleus to the outermost electron—to be much longer than carbon's.

A longer atomic radius usually leads to longer bonds, which are also weaker due to the increased distance. A carbon hydrogen bond is on average 39 kJ/mol stronger than a silicon-hydrogen bond. A carbon-silicon bond is on average 156 kJ/mol weaker than carbon-carbon bonds. For this reason, although silicon is capable of forming similar bonds as carbon, it will likely lose them in favor of other bonds. Silicon will never be stable enough to support life. Carbon-carbon bonds are stronger than silicon-silicon bonds on average. Silicon-silicon bonds break in the presence of UV light, which means such molecules would need to stay out of the sun to survive, thus preventing them from capturing one of the most fundamental sources of Gibbs free energy. Size also limits silicon's ability to form double and triple bonds, further weakening its ability to make bonds strong enough to form the stochastic basis of more-complex molecules.

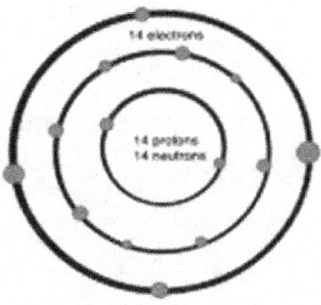

Figure 13.3 Orbitals of a silicon atom

But not all silicon bonds are weaker—in fact, the silicon-oxygen bond is 271 kJ/mol stronger than the carbon oxygen bond. Unfortunately, this bond is so stable that silicon often prefers to bond with oxygen, and once bonded, refuses to bond with much else. It is this bond that makes silicone so stable—and unreactive. While stable in this sense, silicon is too weakly reactive with hydrogen, carbon, and itself to make stable enough molecules to build complex structures. Life must be stable enough not to decompose, but reactive enough to carry out the chemical reactions necessary to build upon. For life, carbon is a perfect fit and silicon is simply not a chemically viable alternative.

Some people claim that with time, anything can happen, and that silicon may form a chemical composition able to support life. But over such a vast timescale, who knows what carbon-based life may have evolved into? Silicon-based life as described by A.G. Cairns-Smith may exist and be evolving. But it would have to be evolving over such a long period of time that rocks may have developed emotions and black holes may have evolved and exploded into galaxies of stars. On a much shorter timescale, quarks and other subatomic particles participate in their own evolutionary dance, faster than we could ever perceive it. In this vision of the universe, eloquently described by Everett, Schrodinger, Bohm, and others in the "many worlds" implicit order, everything is possible—not the least of which may be silicon-based life.

13.12 Imagine

We started this book by trying to understand and explain the origins of life and its evolution through 3.5 billion years. The evolution of life was a dance of interacting matter and energy, increasing in energy efficiency by forming information networks. Life has marched onward, transforming into increasingly complex systems. It has been an exciting experience to watch the stochastic process leading toward the development and evolution of complex adaptive systems. Inevitably, these systems led to the emergence of increasingly complex organisms and eventually to superorganisms. We observed that at each step in the emergence of complexity, living organisms were selected for based on their energy efficiency, often manifested through a lower basal metabolic rate (BMR) per gram. These organisms and superorganisms emerged throughout the progression of evolution, informed by the First and Second Laws of Thermodynamics. These laws, in turn, were clearly evident through efficient signaling within communication networks, which made lower BMR/g possible in more-complex species. Having begun this process, we gradually shifted our attention to the observation that communication and cooperation, from molecule to man, forms the basis for more-efficient, complex organisms and superorganisms to evolve and persist.

We predicted that the interactive computerized algorithmic network (ICAN) that is now evolving is a continuation of that process and will no doubt increase the energy efficiency of our global community. After all, it is informed and instructed by the "useful" and "not useful" input of all its constituents. Responses to these inputs execute the will of the human superorganism based on individual actions, interpreting the collective will of the global community. The ICAN system will use all human and natural resources efficiently and distribute the products and services as dictated by the necessary input of the global community. In fact, such a dynamic system will be a more reliable and stable means of sustenance than any collection of banks or idealistic leaders. The people will be liberated from financial and physical insecurities, as well as unpleasant occupations. Our silicon or carbon-based robots will do the unpleasant tasks for us. As humans, we will be able to pursue what we enjoy, while contributing to the stability of the global community with inner peace and tranquility. We will be rewarded with credit, which we may spend as we wish or gift or loan it to others—with no interest, of course. Unspent credit will be fed back into the system. In addition to credit, scores based on reputation will be given to individuals to match with their contributions to the superorganism. The future world has no need for banks, interest, stock markets, a military, attorneys, accountants, the CIA, or the IRS. Eventually, we will achieve a global "disintermediation," leading to efficient, exponential, and nearly friction-free growth in creativity and innovation. In his book, *The Singularity is Near,* Kurzweil also predicts that disintermediation will continue to transform all industries, as customers are able to interact more directly with products and services without having to go through a middleman.[327]

Physical money will be a relic of the past, replaced by an individual's credit for contributions made to society. In short, global human society will function as a superorganism with over seven billion human individuals interacting and cooperating, like cells in the human body. Our world will become more complex and, thus, more-energy-efficient.

On December 12, 2011, BBC News ran an article with the headline "Unemployment Is World's Fastest-Rising Fear." And it is true that structural unemployment is a natural consequence of rapid innovations in robotics, computerization, networking, and nanotechnology. Unemployment will continue to increase in the near future. This may forge a tumultuous path, as humanity continues toward the critical point of its next emergence. However, in the long run we have nothing to fear. We may come to celebrate the energy efficient byproduct of global human progress. Agricultural civilization resulted in more free time and consequently increased innovation and wealth, and it naturally was followed by the Industrial Revolution, which further increased leisure activity and time to innovate. In the same way, a computerized and information-networking civilization will see more work done by robots and computer-based automation that increases leisure time for people who used to do the work of the machines. There is nothing written in stone that people must work 40 hours per week to deserve a living. In fact, that idea comes from the capitalist paradigm that is designed to siphon people's productivity into the coffers of a few people at the top. This relic of a capitalistic system is the source of unnecessary human suffering and it will be replaced by the ICAN system predicted here. With its result being the more equitable distribution of wealth and free time, the ICAN system will result in a more energy-efficient, happy, and creative life for all people.[328]

To paraphrase the insight of a friend on the nature of our future affluent world: Each individual may have "less stuff, but more time and love." In fact, there will be no fear of poverty and by extension no reason for greed, as money with interest will no longer exist (fulfilling Christ's wishes of 2000 years ago). Everyone will be wealthy in goods, services, and free time, thanks to our AI friends that joined our interactive computerized algorithmic network (ICAN).

This rosy vision may raise some objections. Who will do the dirty work? One answer to this problem is that with the rise of robotics and automated production, the most dangerous, grueling, and difficult human jobs will be passed onto machines. Already, robots explore planets that humans do not have access to. They clean pipes too difficult for us to reach. And they even perform laparoscopic surgeries. Some robots operate aircraft and military robots deactivate bombs. Google has already sent self-driving cars along the coastal highways of California. Using AI software, these cars automatically make decisions about acceleration, braking, turning, and the like, based on an array of data generated by sensors able to detect changes in the speed of other drivers.

I predict that the future of transportation will see an explosion in automation, in ways that transcend our current reliance on rails and roads. In the future, we may see self-guided gyrocopters with one to four passengers, with sensors and software similar to Google cars, enabling them to navigate urban skies across multiple lanes and elevations. In short, automated vehicles will make transportation fluid and relieve congestion, without the fallible input of human drivers or pilots. The problem of congestion in our cities will be resolved, giving us more time to pursue creative projects and activities.

In Steven Johnson's book, entitled *Where Good Ideas Come From*, he noted that the rate of innovation, as related to the size of the community, increases proportionately with its size at an accelerating rate. For example, innovation and invention per capita in New York City scales exponentially with increases in population, dwarfing the creative potential of a small town.[329] The ICAN system that I predict will connect all human brains and make instant communication commonplace within the entire global human community, thus increasing the rate of innovation and invention by leaps and bounds. This is especially true because the credit and reputation of individuals, based on their accomplishments, will be instantly registered and visible to all. There will be no need to hide information with patents or for proprietary reasons. At the present time, we all know that keeping inventions and ideas secret within corporations and by individuals is commonplace and reinforced by patent laws. This is a major impediment to collaborative inventions and innovations. With this impediment out of the way, our future ICAN system will have a very high rate of innovation, as we have seen in the explosion of innovations in computer programming. This is true, despite there being no easy way to compensate or reward the contributor.

We may also wonder what will happen to the criminal, insane, lazy, or antisocial members among us, in the future society described above. The answer is very simple. Outliers of any species are part of the evolutionary process, which is

365

stochastic and selective. All these outliers, including on the other extreme the most beautiful, talented, and intelligent, are part of the necessary Darwinian variability of the gene pool that makes evolution not only possible, but progressive in its path toward greater complexity and efficiency. As humanity moves towards this next emergence, the global ICAN network would have developed a collective moral code that provides compassionate, constructive care for outliers, whether ill or insane, genius or nonconformist. After all, our present moral codes have arisen through collective cultural filters (attractors), i.e., moral codes of different cultures.

As evolution defines its own course, individual humans may be greatly affected. Clearly, the overall global human network, through its interactive information facility, will include a certain percentage of outliers that are necessary parts of any evolutionary process. Biology has operated in this way for billions of years. Ultimately, the next plateau of emergent tranquility in our species will most likely make another jump into a new, emergent state as in punctuated equilibrium as described by Gould. The exact nature of this "new emergence," as we know, will be unpredictable. But this is how it is. There is still room for our collective imagination to cook up new stories, like futurist singularities or religious notions of Eden.

We have seen throughout this book that cooperation and compassion are evolutionarily derived characteristics and are not bestowed by some top-down dictation from the angels above. By

366

extension, we can see that global human morality has also evolved from the bottom-up and is being woven into the fabric of the emerging, interactive global information network to create the global brain. Yes, not only does our emerging network have an energy efficient collective brain, but it also has a compassionate heart. Why? Because it has evolved that way through us. Its structure has determined our survival. In fact, our compassion is a representation of our collective consciousness.

Here, I would like to take the liberty to side with the great 18th century French philosopher, mathematician, and satirist, Voltaire, as he expressed himself through the character Pangloss in his book, *Candide*. Throughout the book, Pangloss repeatedly said, "All is for the best in the best of all possible worlds." Voltaire may be wearing his satirist's smirk but, of course, I take him at his word, which contains many dimensions and layers. Pangloss may have been hanged halfway through the story (and as the future of our complex society unfolds, I may well follow his fate), but I believe this vision, which is in the eye of the beholder, will hold for the continued evolution of our cosmos and species.

Index

References

Introduction

[1] Avery, J. (2003). *Information theory and evolution.* Singapore: World Scientific.
[2] Schneider, E.D., & Sagan, D. (2005). *Into the cool: Energy flow, thermodynamics, and life.* Chicago: University of Chicago Press.
[3] Solé, R., & Goodwin, B. (2000). *Signs of life: How complexity pervades biology.*

Chapter 1

[4] 11 awesome facts about lightning, 09-26-13 by Kids Discover
[5] Tom Seigfried (2010), Editor in Chief for Science News, made this statement in October "from the editor" article

Chapter 3

[6]   shttp://insidemathblog.blogspot.com/2015/09/polyhedron.html Inside Math Blog Sept 7, 2015

Chapter 4

[7] Source: http://anthro.palomar.edu/synthetic/synth_9.htm Phyletic Gradualism vs Punctuated Equilibrium https://is.theorizeit.org/wiki/Punctuated_equilibrium_theory

Chapter 5

[8] *Figure 5.1* – The Lorenz attractor; from Mosekilde et al. (1994)
[9] Martini, F., and Judi, N. (2009). *Fundamentals of anatomy & physiology.* San Francisco: Pearson Education.
[10] Duckworth, W.C., Bennett R.G., & Hamel F.G. (1998). Insulin degradation: progress and potential, *Endocrine Reviews*, 19, 608–624.

Chapter 6

[11] *Figure 6.2* – Dr. Joan Adler, Physics, Technion-Israel Institute of Technology. http://phycomp.technion.ac.il/~phr76ja/lecture1.html

[12] *Figure 6.3* – Bio+Chem Notes. ^-^. http://as-bio-and-chem.blogspot.com/2010/09/recapping-rates-of-reactionkinetics.html

[13] *Figure 6.5* – Professor Olsen @ Large (blog). http://diogenesii.wordpress.com/2013/03/07/march-7-1930-a/

[14] Kauffman, S.A. (1993). *The origins of order: Self-organization and selection in evolution.* New York: Oxford University Press.

[15] Hazen, R. (2005). Genesis: The scientific quest for life's origins. Joseph Henry Press.

[16] Figure 6.6 – Frances J. Sharom Ph.D. http://www.uoguelph.ca/~fsharom/research/research.shtml

[17] *Figure 6.7* – Open Curriculum http://www.theopencurriculum.org/articles/biology/?q=earlylife&revision=171

[18] Johnston, W., et al. (2001). RNA-catalyzed RNA polymerization: accurate and general RNA-templated primer extension. *Science, 292,* 1319-25.

[19] P.J., U., & Bartel D.P. (1998). RNA-catalysed nucleotide synthesis. *Nature, 395,* 260–263.

[20] Zhang, B, & Cech, T.R. (1997). Peptide bond formation by in vitro selected ribozymes. *Nature, 390,* 96–100.

[21] Hazen, R. (2005). Genesis: The scientific quest for life's origins.

[22] Markovitch, O. & Agmon, N. (2007). Structure and energetics of the hydronium hydration shells. *Journal of Physical Chemistry*, 111, 2253–2256.

[23] Baker, W. Gann, B., & Levine, L. (2008). *Molecular biology of the gene.* Pearson Education.

[24] Vollhardt, S. (2007). *Organic chemistry structure and function* (5th ed.). W. H. Freedman and Company.

[25] Watson, J.D., et al. (2008). *Molecular biology of the gene.* Pearson Education.

[26] Pico, R.M. (2001). *Consciousness in four dimensions: biological relativity and the origins of thought* (1st ed.). New York: McGraw-Hill.

Chapter 7

[27] Slonczewski J.L., & Foster, J.W. (2010). *Microbiology: An evolving science*. W. W. Norton & Company.

[28] Searls D.B. (2002). The language of genes, *Nature, 420*, 211-217.

[29] Long, D.M., & Uhlenbeck, O.C. (1993). Self-cleaving catalytic RNA. *FASEB Journal, 7*, 25-30.

[30] Bassler B.L. (2002). Small talk: cell-to-cell communication in bacteria. *Cell Biology*, 109, 421-424.

[31] *Figure 3.1* - http://openwetware.org/wiki/CH391L/S13/QuorumSensing

[32] Ben-Jacob E., et al. (1998). Cooperative organization of bacterial colonies: from genotype to morphotype. *Annual Review of Microbiology, 52*, 779-806.

[33] Kozlovsky, Y., Cohen, I., Golding, I., & Ben-Jacob E. (1999). Lubricating bacteria model for branching growth of bacterial colonies. *Physics Review, E59*, 7025-7035.

[34] *Figure 7.2* - http://en.wikipedia.org/wiki/File:Paenibacillus_dendritiformis_colony.png

[35] Rosenberg, E., & Varon M. (1984). Antibiotics and lytic enzymes. In M. Rosenberg (Ed.). *Myxobacteria: Development and cell interactions* (pp.109-125). New York: Springer.

[36] Shi, W., Kohler, T., & D.R. Zusman (1993). Chemotaxis plays a role in the social behavior of Myxococcus xanthus. *Molecular Microbiology, 9*, 601-611.

[37] Burnham J.C., Collart S.A., & Highison, B.W. (1981). Entrapment and lysis of the cyanobacterium Phormidium luridum by aque-ous colonies of Myxococcus xanthus PCO2. *Archives of Microbiology, 129*, 285-294.

[38] Fisher, L. (2009). Through worldwide biodiversity inventory slime mold gets respect. *University of Arkansas Research Frontiers*.

[39] Conover, A. (2001). Hunting slime molds. *Smithsonian Magazine*, http://www.smithsonianmag.com/sciencenature/phenom_mar01.html.

[40] Figure 7.3 - Sathiyanarayanan Manivannan, http://www.biotech.iitm.ac.in/faculty/slimemolds.html

[41] Martiel, J.L., & Goldbeter, A. (1987). A model based on receptor desensitization for cyclic AMP signaling in Dictyostelium cells. *Biophysical Journal, 52*, 807-828.

[42] Schaap, P. (2011). Evolution of developmental cyclic adenosine

monophosphate signaling in the Dictyostelia from an amoebozoan stress response. *Development, Growth, & Differentiation,* 53, 452-462.

[43] Wright, B.E. and Bloom, B. (1961). In vivo evidence for metabol-ic shifts in the differentiating slime mold. *Biochimica et Biophysica Acta,* 48, 342-346.

[44] Goodman E.M., & Beck, T. (1974). Metabolism during differenti-ation in the slime mold Physarum polycephalum. *Canadian Journal of Microbiology,* 20, 107-111.

[45] Source for image from http://www.metafysica.nl/dissipative_systems.html.

[46] Wikimedia commons. *Spiral galaxy.*

[47] Wilson, E.O. (2000). *Sociobiology: The new synthesis* (25th ed.). Belknap Press: London.

[48] Bonner, J.T. (1998). The origins of multicellularity. *Integrative biology: Issues, news, and reviews,* 1, 27–36.

Chapter 8

[49] Björkstén, B., et al. (2001). Allergy development and the intestinal microflora during the first year of life. *Journal of Allergy and Clinical Immunology,* 108, 516–20.

[50] Guarner, F., & Malagelada, J.R. (2003). Gut flora in health and disease. *Lancet,* 361, 512–519.

[51] Sears, C.L. (2005). A dynamic partnership: Celebrating our gut flora. *Anaerobe,* 11, 247–251.

[52] Steinhoff, U. (2005). Who controls the crowd? New findings and old questions about the intestinal microflora. *Immunology. Letter,* 99, 12–16.

[53] Törnroth-Horsefield, S., & Neutze, R. (2008). Opening and clos-ing the metabolite gate. *Proceedings of the National Academy of Sciences,* 105, 19565–19566.

[54] *Figure 8.1* – Detroit Medical Center, http://www.dmc.org/information-for-mitochondrial-medicine-centerpatients.html

[55] Mereschkowsky, C. (1905). Über Natur und ursprung der chroma-tophoren im pflanzenreiche. *Biol Centralbl,* 25, 593–604.

[56] Margulis, L. (1970). *Origin of Eukaryotic Cells.* New Haven: Yale University Press.

[57] Keeling, P., Leander B., & Simpson A., (2009). Eukaryota, organisms with nucleated cells. *Tree of Life Web Project.*

[58] Inoue, K. (2007). The chloroplast outer envelope membrane: the edge of light and excitement. *Journal of Integrative Plant Biology*, 49, 1100–1111.

[59] Holtfreter, J. (1939). Gewebeaffinitat: ein mittel der embryonalen formbildung. Arch Exp Zellforsch Besonders Gewebezucht, 23, 169-209. Revised and reprinted in English in Willier, B.H., & Oppenheimer, J. M. (Eds.) (1964), *Foundations of experimental embryology* (pp.186–225). Englewood Cliffs: Prentice-Hall.

[60] Müller, G., & Newman, S. (2003). *Origination of organismal form: Beyond the gene in developmental and evolutionary biology.* Cambridge: The MIT Press.

[61] Burnet F.M., (1960). *Immunological recognition of self.* Nobel Lecture.

[62] Christensen, B. (2006). Mind control by Parasites. *Live Science.* http://www.livescience.com/7019-mind-control-parasites.html.

[63] Boesch, C., et al. (2005). *Animal Social Complexity: Intelligence, Culture, and Individualized Societies.* Cambridge: Harvard University Press.

[64] Eknoyan, G. (1999). Santorio sanctorius (1561–1636) - founding father of metabolic balance studies. *American Journal of Nephrology*, 19, 226–33.

[65] Animal Aging and Longevity Database. *Human Aging Genomic Resources.* http://genomics.senescence.info/species/.

[66] DeLong J.P. et al. (2010). Shifts in metabolic scaling, production, and efficiency across major evolutionary transitions of life. *Proceedings of the National Academy of Sciences*, 107, 1294112945.

[67] DeLong J.P. et al. (2010). Shifts in metabolic scaling, production, and efficiency across major evolutionary transitions of life.

[68] West, G.B., Brown, J.H., & Enquist B.J. (1997). A general model for the origin of allometric scaling laws in biology. *Science*, 276, 122–126.

[69] Makarieva, A.M, Gorshkov V.G, & Li, B. (2005). Energetics of the smallest: Do bacteria breathe at the same rate as whales? *Proceedings of the Royal Society B: Biological Sciences*, 272, 2219-2224.

[70] Stal, L., & Moezelaar, R. (1997). Fermentation in cyanobacteria. *FEMS Microbiology Reviews*, 21, 179-211.

[71] Brown, West. (1997) *Scaling in Biology.* Oxford University press, Inc.

[72] John P. DeLonga,, Jordan G. Okiea, Melanie E. Mosesa, Richard M. Siblyd, and James H. Brown (2010) "Shifts in metabolic scaling,

production, and efficiency across major evolutionary transitions in life." *PNAS* vol. 107 (29) pages 12941-12945

[73] Stumpf, Porter (2012) "The Critical Truth about power Laws" *Science*. Vol. 335, page 665-666.

[74] Raven, Peter H. (2011). *Biology: ninth edition*. New York: McGraw-Hill. pp. 278–301

[75] Mohamed Ouhammouch, Robert E. Dewhurst, Winfried Hausner, Michael Thomm, and E. Peter Geiduschek (2003); "Activation of archaeal transcription by recruitment of the TATA-binding protein". *PNAS* Col. 100 (9): page 5097– 5102.

[76] Sara Moens and J. Vanderleyden. (1997) "Glycoproteins in prokaryotes." Archives of Microbiology. Volume 168(3), page 169-175.

[77] Alon. (2007) An Introduction to systems Biology: *Design Principles of Biological Circuits*. Taylor and Francis group, LLC. Page 34

[78] Thomas Pfeiffer, Stefan Schuster, Sebastian Bonhoeffe. (2001) "Cooperation and Competition in the Evolution of ATP-Producing Pathway." *Science*. Vol. 292, page 504-507

[79] Alan E. Senior, , Sashi Nadanaciva, Joachim Weber. (2002) "The molecular mechanism of ATP synthesis by F1F0-ATP synthase." *Bioenergetics*.Vol 1553(3), Pages 188–211

[80] Michael Lynch and John S. Conery (2003) "The Origins of Genome Complexity" *Science* Vol. 302, page 1401-1404

[81] A. K. Saha, T. G. Kurowski, and N. B. Ruderman (1995) "A malonyl-CoA fuel-sensing mechanism in muscle: effects of insulin, glucose, and denervation." *AJP - Endo* vol. 269(2) pages E283-E289

[82] Behzad Mohit, Thermoinfocomplexity: A Comprehensive Theory of Evolution, 2012.

Chapter 9

[83] Goodson, J.L., et al. (2009). Mesotocin and nonapeptide receptors promote estrildid flocking behavior. *Science*, 325, 862-866.

[84] Lissaman, P., & Shollenberger, C. (1970). Formation flight of birds. *Science*, 168, 1003-1005.

[85] Weimerskirch, H., et al., (2001). Energy saving in flight formation: pelicans flying in a 'V' can glide for extended periods using other birds' airstreams. *Nature*, 413, 697.

[86] Photograph by Julo. Wikimedia commons.

[87] Potts, W.K. (1984). The chorus-line hypothesis of coordination in avian flocks. *Nature*, 24, 344-345.

[88] Shaw, E. (1978). Schooling fishes. *American Scientist*, 66, 166–175.

[89] Waloff, Z. (1966). The upsurges and recessions of the desert lo-cust plague: an historical survey. *London: Anti-Locust Research Centre.*

[90] Milne, L., & Milne, M. (1992). *The Audubon Society field guide to North American insects and spiders*. New York: Alfred A Knopf.

[91] Dawkins, R. (1986). *The blind watchmaker*. New York: Norton & Company.

[92] Abrahams, M. & Colgan, P. (1985). Risk of predation, hydrody-namic efficiency, and their influence on school structure. *Environmental Biology of Fishes*, 13, 195-202.

[93] Queiroz, H., & Magurran, A.E. (2005). Safety in numbers? Shoal-ing behavior of the Amazonian red-bellied piranha, *Biology Letters*, 1, 155-157.

[94] Sword, G.A. Lorch, P.D., & Gwynn, D.T. (2005). Insect behavior: Migratory bands give crickets protection. *Nature*, 433, 703.

[95] Roberts, G. (1996). Why individual vigilance increases as group size increases. *Animal Behavior*, 51, 1077-1086.

[96] Lima, S. (1995). Back to the basics of anti-predatory vigilance: The group-size effect. *Animal Behavior*, 49, 11-20.

[97] Kils, U. (1992). The ecoSCOPE and dynIMAGE: Microscale tools for in situ studies of predator-prey interactions. *Archives Hydrobiology Beih*, 36, 83-96.

[98] Partridge, B., Johansson, J., & Kalish, J. (1983). The structure of schools of giant bluefin tuna in Cape Cod Bay. *Environmental Biology of Fishes*, 9, 253.

[99] Pitcher and Parish 1993.

Chapter 10

[100] Darwin, C. (2003). *On the origin of species: A facsimile of the first edition*. Rockville: Wildside Press.

[101] Darwin, C. (2003). *On the origin of species: A facsimile of the first edition*.

[102] Wilson, E.O. (2000). *Sociobiology: The new synthesis*. Cambridge: The Belknap Press of Harvard University Press.

[103] Darwin, C. (2003). *On the origin of species: A facsimile of the first edition*.

[104] Morgan, R.C. (2004). *Biology, husbandry and display of the diurnal honey ant Myrmecocystus mendax Wheeler (Hymenoptera: Formicidae).* Insectarium, Cincinnati Zoo and Botanical Gardens.

[105] Honeypot ants. Wikimedia commons.

[106] Wilson, E.O. (2000). *Sociobiology: The new synthesis.*

[107] Dawkins R. (1990). *The selfish gene.* New York: Oxford University Press.

[108] Hamilton, W.D. (1964). The genetical evolution of social behaviour I and II. *Journal of Theoretical Biology*, 7, 1-16 and 17-52.

[109] Keim, B. (2010) E.O. Wilson Proposes New Theory of Social Evolution. *Wired.* http://www.wired.com/wiredscience /2010/08/kin-selection-challenged/

[110] Darwin, C. (2003). *On the origin of species: A facsimile of the first edition.*

[111] Rabeling C., Brown, J.M., & Verhaagh M. (2008). Newly discovered sister lineage sheds light on early ant evolution. *Proceedings form the National Academy of Sciences*, 105, 14913–14917.

[112] Schultz, T. (2000). In search of ant ancestors. *Proceedings of the National Academy of Sciences*, 97, 14028–14029.

[113] Ndabahaliye, A. (2002). Number of neurons in the human brain. *The Physics Factbook.*

[114] Vogl, W., et al. (2002). Cuckoo females preferentially use specific habitats when searching for hot nests. *Animal Behavior*, 64, 843–850.

[115] Peer, B., Robinson, S., & Herkert, J. (2000). Egg rejection by cowbird hosts in grasslands. *The Auk*, 117, 892–901.

[116] James, S., et al. (2010). Allometric scaling of metabolism, growth, and activity in whole colonies of the seed-harvester ant Pogonomyrex californicus.

[117] Hou, C., et al. (2010). Energetic basis of colonial living in social insects. *Proceedings of the National Academy of Sciences*, 107, 3634-3638.

[118] Soutwick, E.E., & Heldmaier, G. (1987). Temperature control in honey bee colonies. *Bioscience*, 37, 395-399.

[119] Vollmer, S.V., & Edmunds, P.J. (2000). Allometric scaling in small colonies of the Scleractinian Coral Siderastrea siderea. *Biology Bulliten*, 199, 21-28.

[120] Nakaya, F., Saito, Y., & Motokawa, T. (2003). Switching of metabolic rate scaling between allometry and isometry in colonial ascidians. *Proceedings of the Royal Society B: Biological Sciences*,

270, 1105-1113.

[121] James, S., et al. (2010). Allometric scaling of metabolism, growth, and activity in whole colonies of the seed-harvester ant Pogonomyrex californicus.

[122] Wade, N. (2008). Taking a cue from ants on evolution of humans. *New York Times.* http://www.nytimes.com/2008/07/15/science/15wils.html.

[123] Schultz, T.R. (2000). In search of ant ancestors. *Proceedings of the National Academy of Sciences*, 97, 14028–14029.

[124] Wehner, R., Fukushi, T, & Isler, K. (2007). On being small: Brain allometry in ants. *Brain Behavior and Evolution*, 69, 220-8.

Chapter 11

[125] Wilkinson, G.S., (1984). Reciprocal food sharing in the vampire bat. *Nature*, 308, 181–184.

[126] Wilkinson, G.S., (1990). Food sharing in vampire bats. *Scientific American*, 76–82.

[127] Schmidt, U. (1972). Die sozialen laute juveniler vampirfledermäuse (Desmodus rotundus) und ihrer mütter" (Social calls of juvenile vampire bats (D. rotundus) and their mothers), *Bonner Zoologische Beitreage*, 23, 310–316.

[128] Wilkinson, G.S., (1986). Social grooming in the common vampire bat, Desmodus rotundus. *Animal Behavior*, 34, 1880-1889.

[129] Wilkinson, G.S., (1984). Reciprocal food sharing in the vampire bat.

[130] Sayigh, L.S., et al., (1999). Individual recognition in wild bottle-nose dolphins: A field test using playback experiments. *Animal Behaviour*, 57, 41–50.

[131] Wells, R.S. (2000). Reproduction in wild bottlenose dolphins: Overview of patterns observed during a long-term study. In D. Duffield, & T. Robeck (Eds.), *Bottlenose dolphin reproduction workshop report* (pp. 57–74). Silver Springs: AZA Marine Mammal Taxon Advisory Group.

[132] Whiten, A., et al. (1999). Cultures in chimpanzees. *Nature*, 399, 682–685.

[133] Smolker, R., et al. (1997). Sponge carrying by Indian Ocean bottlenose dolphins: Possible tool use by a delphinid. *Ethology*, 103, 454–465.

[134] Wells, R.S., & Scott, M.D. (1987). The social structure of freeranging bottlenose dolphins. *Current Mammology*, 1, pp. 247–305.

[135] Nowacek, D.P. (1999). *Sound use, sequential behavior and foraging ecology of foraging bottlenose dolphins, Tursiops truncatus* (Doctoral dissertation). Retrieved from Massachusetts Institute of Technology and Woods Hole Oceanographic Institution.

[136] Nowacek, D.P. (1999). Sound use, sequential behavior and forag-ing ecology of foraging bottlenose dolphins, Tursiops truncatus.

[137] Moors, T.L. (1997). *Is 'menage a trois' important in dolphin mating systems? Behavioral patterns of breeding female bottlenose dolphins.* (Master thesis). Retrieved from University of California, Santa Cruz.

[138] U.S. Department of Commerce, National Oceanic and Atmos-pheric Administration, National Marine Fisheries Service. (1993). Coastal stock(s) of Atlantic bottlenose dolphin: Status review and management. *Proceedings and Recommendations from a Workshop held in Beaufort, North Carolina, US Department of Commerce, National Oceanic and Atmospheric Administration, National Marine Fisheries Service*, 56–57.

[139] Lilley, R. (2008). Dolphin saves stuck whales, guides them back to sea. *National Geographic News.*

[140] Free, C. (2008). Shark! How one surfer survived an attack. *Reader's Digest.* http://www.rd.com/true-stories/survival/sharkattack-dolphins-save-surfer-from-shark/

[141] Simões-Lopes, P.C. (1991). Interaction of coastal populations of Tursiops truncates (Cetacea, Delphinidae) with the mullet artisanal fisheries in southern Brazil. *Biotemas*, 4, 83–94.

[142] Peterson, D., Hanazaki, N., & Simões-Lopes, P.C. (2008). Natural resource appropriation in cooperative artisanal fishing between fisherman and dolphins (Tursiops truncatus) in Laguna, Brazil. *Ocean & Coastal Management*, 51, 469-475.

[143] Moberg, K.U., & Francis R. (2003). *The oxytocin factor: Tapping the hormone of calm, love, and healing.* Cambridge: Da Capo Press.

[144] Pert, C. (1999). *Molecules of emotion: The science behind mindbody medicine.* New York: Simon & Schuster.

[145] Rao, P.D.P., & Kanwal, J.S. (2004). Oxytocin and vasopressin immunoreactivity within the forebrain and limbic-related areas in the mustached bat, Pteronotus parnellii. *Brain Behavior and Evolution*, 63, 151-168.

[146] Goodson, J.L. (2009). Mesotocin and nonapeptide receptors pro-mote Estrildid flocking behavior. *Science*, 325, 862-866.

[147] Insel, T.R. (2010). The challenge of translation in social neuroscience: a review of oxytocin, vasopressin, and affiliative

behavior. *Neuron*, 65, 768-779.

[148] Levine A., et al. (2007). Oxytocin during pregnancy and early postpartum: individual patterns and maternal-fetus attachment. *Peptides*, 28, 1162-1169.

[149] Levine A., et al. (2007). Oxytocin during pregnancy and early postpartum: individual patterns and maternal-fetus attachment. *Peptides*, 28, 1162-1169.

Chapter 12

[150] Johnson, A.W., & Earle, T. (2000). *The evolution of human societies: From foraging group to agrarian state.*

[151] For a more detailed discussion of the history of theories of sociocultural evolution, see Johnson, A.W., & Earle, T. (2000). *The evolution of human societies: From foraging group to agrarian state*, 2.

[152] Read, D.W., & LeBlanc, S.A. (2003). Population growth, carrying capacity, and conflict. *Current Anthropology*, 44, 59-85.

[153] Keeley, L.H. (1988). Hunter-Gatherer economic complexity and "population pressure": A cross-cultural analysis. *Journal of Anthropological Archaeology*, 7, 373-411.

[154] Burnside, W.R. et al. (2012). Human macroecology: Linking pattern and process in big-picture human ecology. *Biology Review*, 87, 194-208.

[155] Nettle, D. (2009). Ecological influences on human behavioral diversity: a review of recent findings. *Trends in Ecology & Evolution*, 24, 618-624.

[156] Nakagaki, T. et al. (2000). Maze-solving by an amoeboid organism. *Nature*, 407, 470.

[157] Nakagaki, T. et al. (2000). Maze-solving by an amoeboid organism.

[158] Nakagaki, T. et al. (2000). Maze-solving by an amoeboid organism.

[159] *Figure 12.1* – Public Domain by author Philcha, http://commons.wikimedia.org/wiki/File:Slime_mold_solves_maze.png

[160] Li, S.I., & Purugganan, M.D. (2011). The cooperative amoeba: Dictyostelium as a model for social evolution. *Trends in Genetics*, 27, 48-54.

[161] Kaplan, H., et al. (2000). A theory of human life history evolu-tion: Diet, intelligence, and longevity. *Evolutionary Anthropology*, 9, 156-185.

[162] Johnson, A.W., & Earle, T. (2000). The evolution of human

societies: From foraging group to agrarian state.

[163] Burnside, W.R., et. al. (2012). Human macroecology: Linking pattern and process in big-picture human ecology.

[164] Hamilton, M.J., et al. (2007) The complex structure of hunter-gatherer social networks. Proceedings of the Royal Society B, 274, 2195-2202.

[165] Burnside, W.R. et al. (2012). Human macroecology: Linking pattern and process in big-picture human ecology.

[166] Johnson, A.W., & Earle, T. (2000). *The evolution of human societies: From foraging group to agrarian state.*

[167] Binford, L. (1980). Willow smoke and dogs' tails: Hunter-gatherer settlement systems and archaeological site formation. *American Antiquity*, 45, 4-20.

[168] Burnside, W.R. et. al. (2012). Human macroecology: Linking pattern and process in big-picture human ecology.

[169] Johnson, A.W., & Earle, T. (2000). *The evolution of human societies: From foraging group to agrarian state*, 46.

[170] Hamilton, M.J., et al. (2007). The complex structure of hunter-gatherer social networks.

[171] Burnside, W.R. et. al. (2012). Human macroecology: Linking pattern and process in big-picture human ecology.

[172] Lee, R. (1979).*The !Kung San.* Cambridge: Cambridge University Press.

[173] Johnson, A.W., & Earle, T. (2000). *The evolution of human societies: From foraging group to agrarian state*, 70.

[174] Johnson, A.W., & Earle, T. (2000). *The evolution of human societies: From foraging group to agrarian state*, 67.

[175] Johnson, A.W., & Earle, T. (2000). *The evolution of human societies: From foraging group to agrarian state*, 67.

[176] Binford, L. (1980). Willow smoke and dogs' tails: Hunter-gatherer settlement systems and archaeological site formation. *American Antiquity*.

[177] Burnside, W.R. et. al. (2012). Human macroecology: Linking pattern and process in big-picture human ecology.

[178] Hamilton, M. et al. (2009). Population stability, cooperation, and the invasibility of the human species. *Proceedings from the National Academy of Sciences*, 106, 12255-12260.

[179] Hamilton, M. et al. (2007). Nonlinear scaling of space use in hu-man hunter-gatherers. *Proceedings from the National Academy of Sciences*, 104, 4765-4769.

[180] Hamilton, M. et al. (2009). Population stability, cooperation, and the invasibility of the human species.

[181] Hamilton, M. et al. (2007). Nonlinear scaling of space use in hu-man hunter-gatherers.

[182] Burnside, W.R. et. al. (2012). Human macroecology: linking pattern and process in big-picture human ecology, 198.

[183] Hamilton, M. et al. (2007). Nonlinear scaling of space use in hu-man hunter-gatherers.

[184] Johnson, A.W., & Earle, T. (2000). *The evolution of human societies: From foraging group to agrarian state*, 71.

[185] Source for *Figure 12.4*: Sahlins, M.D. (1972). *Stone age economics*. Piscataway: Transaction Publishers.

[186] Johnson, A.W., & Earle, T. (2000). *The evolution of human societies: From foraging group to agrarian state*, 48.

[187] Johnson, A.W., & Earle, T. (2000). *The evolution of human societies: From foraging group to agrarian state*, 44.

[188] Johnson, A.W., & Earle, T. (2000). *The evolution of human societies: From foraging group to agrarian state*, 44.

[189] Hamilton, M. et al. (2007). Nonlinear scaling of space use in human hunter-gatherers.

[190] Burnside, W.R. et al (2012). Human macroecology: Linking pattern and process in big-picture human ecology.

[191] Johnson, A.W., & Earle, T. (2000). *The evolution of human societies: From foraging group to agrarian state*, 61.

[192] Read, D.W., & LeBlanc, S.A. (2003). Population growth, carrying capacity, and conflict.

[193] Keeley, L.H. (1988). Hunter-Gatherer economic complexity and "population pressure": A cross-cultural analysis.

[194] Richerson, P.J., et al. (2001). Was agriculture impossible during the Plestocene but mandatory during the Holocene? A climate change hypothesis. *Society for American Archaeology*, 66, 387411.

[195] Richerson, P.J., et al. (2001). Was agriculture impossible during the Plestocene but mandatory during the Holocene? A climate change hypothesis.

[196] Clark, P.E., et al. (1999). Northern hemisphere ice-sheet influences on global climate change. *Science*, 286, 1104-1111.

[197] Dansgaard, W., et al. (1993). Evidence for general instability of past climate from a 250-kyr ice-core record. *Nature*, 364, 218220.

[198] Ditlevsen, P.D., et al. (1996). Contrasting atmospheric and cli-mate dynamics of the last-glacial and Holocene periods. *Nature*, 379, 810-812.

[199] GRIP (Greenland Ice-core Project Members) (1993). Climate instability during the last interglacial period recorded in the GRIP Ice Core. *Nature*, 364, 203-207.

[200] Richerson, P.J., et al. (2001). Was agriculture impossible during the Plestocene but mandatory during the Holocene? A climate change hypothesis.

[201] Tellier, L.N. (2009). *Urban world history: An economic and geographical perspective.* Quebec: University of Quebec Press.

[202] Kenneth, W.H. (1998). *Population estimates of the Roman Empire.* Tulane University,
http://www.tulane.edu/~august/H303/handouts/Population.htm

[203] History of Europe, Demographic and agricultural growth. *Encyclopedia Britannica.*

[204] Population Division (2010). *UN world population prospects, 2010 revision.* United Nations Department of Economics and Social Affairs.

[205] Population Division (2010). *UN world population prospects, 2010 revision.*

[206] Population Division (2004). *World population to 2300.* United Nations Department of Economic and Social Affairs.

[207] U.S. Census Bureau. (2013). *World POPClock Projection,*
http://www.census.gov/population/popclockworld.html.

[208] U.S. Census Bureau. *Wikimedia Commons.*

[209] Fuller, D.Q., Willcox, G., & Allaby, R.G. (2011). Cultivation and domestication had multiple origins: Arguments against the core area hypothesis for the origins of agriculture in the Near East. *World Archeology*, 43, 628-652.

[210] Bar-Yosef, O. (1998). The Natufian culture in the Levant, thresh-old to the origins of agriculture. *Evolutionary Anthropology*, 6, 159-177.

[211] Milstein, M. (2008). Oldest shaman grave found. *National Geographic*, http://news.nationalgeographic.com/news/2008/11/081104-israel-shaman-missions.html.

[212] Burston, B. (2008). Hebrew U. unearths 12,000-year old skeleton of 'petite' Natufian priestess. *Haaretz*, http://www.haaretz.com/print-edition/news/hebrew-u-unearths-12-000-year-old-skeleton-of-petite-natufian-priestess-1.256664

[213] Bar-Yosef, O. (1998). The Natufian culture in the Levant, thresh-old to the origins of agriculture.

[214] Mithen, S. (2006). *After the ice: A global human history, 20,005,000 BC.* Cambridge: Harvard University Press, 59.

[215] Liran, R., & Barkal, R. (2011). Casting a shadow on Neolithic Jericho. *Journal of Antiquity,* 85.

[216] Negev, A., & Gibson, S. (2005). *Archeological Encyclopedia of the Holy Land.* New York: Continuum International Publishing Group, 180.

[217] Kuijt, I., & Finlayson, B. (2009). Evidence for food storage and predomestication granaries 11,000 years ago in the Jordan Valley. *Proceedings of the National Academy of Sciences,* 106, 10966.

[218] Kuijt, I., & Finlayson, B. (2009). Evidence for food storage and predomestication granaries 11,000 years ago in the Jordan Valley.

[219] Kuijt, I., & Finlayson, B. (2009). Evidence for food storage and predomestication granaries 11,000 years ago in the Jordan Valley.

[220] Bar-Yosef, O. (1998). The Natufian culture in the Levant, thresh-old to the origins of agriculture.

[221] Grimaldi, D. (2009). Pushing Back Amber Production. *Science,* 326, 51-52.

[222] Beck, C.W. (1985). Criteria for amber trade: The evidence in the eastern European Neolithic. *Journal of Baltic Studies,* 16, 276292.

[223] Penhallurick, R.D. (1986). *Tin in antiquity: Its mining and trade throughout the ancient world with particular reference to Cornwall.* London: The Institute of Metals.

[224] Pulak, C. (2001). The cargo of the Uluburun ship and evidence for trade with the Aegean and beyond. In L. Bonfante, & V. Karageogrhis (Eds.), *Italy and Cyprus in Antiquity: 1500-450 BC* (pp. 12-61). Nicosia: The Costakis and Leto Severis Foundation.

[225] Cierny, J., & Weisgerber, G. (2003). The Bronze Age tin mines in Central Asia. In A. Giumlia-Mair, L. Schiavo (Eds.), *The Problem of Early Tin* (pp.23-31). Oxford: Archaeopress.

[226] Gould, S.J., & Vrba, E.S. (1982). Exaptation: A missing term in the science of form. *Paleobiology,* 8, 4-15.

[227] Meyer, E. (1907). *Geschichte des Altertums IV.* Basel: Stuttgart J.G. Cotta.

[228] Briant, P. (2012) From the Indus to the Mediterranean: The administrative organization and logistics of the great roads of the Achaemenid Empire. In S. Alcock , & J. Bodel (Eds.), *Highways, Byways, and Road Systems in the Pre-Modern World.* Chichester: Wiley-Blackwell.

[229] Wikimedia Commons. http://en.wikipedia.org/wiki/File:Persian_Empire_490_BC.png

[230] Imanpour, M-T. (2010). The communication roads in Parsa dur-ing the Achaemenid Period. In *Ancient and Middle Iranian Studies:*

Proceedings of the 6th European Conference of Iranian Studies, Vienna, 18-22 September 2007, edited by Macuch, M. et al. Wiesbaden: Harrassowitz.

[231] Henkelman, W. (2010) Consumed before the king: The table of Darius, that of Irdabama and Irtastuna, and that of his satrap, Karkis, In B. Jacobs, & R. Rollinger (Eds.), *Der Achamenidenhof (Classica et Orientalia 2)* (pp. 667-775). Wiesbaden: Harrassowitz

[232] Benveniste, E. (1958). Notes sur les tablettes elamites de Persepo-lis. *Journal Asiatique*, 246, 49-65.

[233] Hallock, R.T. (1969). *Persepolis fortification tablets*. Chicago: Oriental Institute Publications.

[234] Briant, P. (2002). *From Cyrus to Alexander: A history of the Persian Empire*. Winona Lake: Eisenbrauns.

[235] Goodspeed, G.S. (1899). The Persian Empire from Darius to Ar-taxerxes. *The Biblical World*, 14, 251-257.

[236] Smith, S. (1944). *Isaiah Chapters XL-LV*. London: Oxford University Press.

[237] Mayrhofer, M. (1964). *Handbuch des Altpersischen*. Wiesbaden: Harrassowitz.

[238] Boyce, M. (1991). *A history of Zoroastrianism*. New York: E.J. Brill.

[239] Boyce, M. Achaemenid religion. *Encyclopedia Iranica,* www.iranicaonline.org

[240] Shahbazi, A.S. (1994). Darius I the Great. *Encyclopedia Iranica*, 7, 41-50.

[241] Schaeder, H.H. (1932). Die ionier in der bauinschrift des dareios von susa. *Jahrbuch des Deutschen Archäologischen Instituts,* 270-74.

[242] Source for *Figure 12.7*: http://en.wikipedia.org/wiki/File:Map_achaemenid_empire_en.png. Wikimedia commons.

[243] Russell, J.C. (1972). Population in Europe. In C. Cipolla (Ed.), *The Fontana economic history of Europe, Vol. I: The Middle Ages*. Glasgow: Collins/Fontana, 25-71

[244] Burnett, C.S.F. (1977). A Group of Arabic-Latin translators working in northern Spain in the mid-12th century. *Journal of the Royal Asiatic Society of Great Britain and Ireland*, 109, 62108.

[245] Majeed, A. (2005). How Islam changed medicine: Arab physi-cians and scholars laid the basis for medical practice in Europe. *British Medical Journal*, 331, 24-31.

[246] El Diwani, R. (2005). Islamic contributions to the West. *Talk given*

at Lake Superior State University, Sault Ste. Marie, MI, as Fulbright visiting specialist.

247 Kaviani, R. et al. (2012). The significance of the Bayt Al-Hikma (House of Wisdom) in early Abbasid Caliphate. *Middle-East Journal of Scientific Research,* 11, 1272-1277.

248 Kaviani, R. et al. (2012). The significance of the Bayt Al-Hikma (House of Wisdom) in early Abbasid Caliphate.

249 Kaviani, R. et al. (2012). The significance of the Bayt Al-Hikma (House of Wisdom) in early Abbasid Caliphate.

250 Ferguson, N. (2004). *Empire: The rise and demise of the British world order and the lessons for global power.* New York: Basic Books.

251 Allen, R. (2009) The British Industrial Revolution in Global Perspective. Cambridge: Cambridge University Press.

252 Allen, R. (2009). *The British Industrial Revolution in global perspective.*

253 Wills, J.E, Jr. (1993). Maritime Asia, 1500-1800: The interactive emergence of European Domination. *The American Historical Review,* 98, 83-105.

254 Wills, J.E, Jr. (1993). Maritime Asia, 1500-1800: The interactive emergence of European Domination.

255 Wills, J.E, Jr. (1993). Maritime Asia, 1500-1800: The interactive emergence of European Domination.

256 Wills, J.E, Jr. (1993). Maritime Asia, 1500-1800: The interactive emergence of European Domination.

257 Bayly, C.A. (1999). *Empire and Information: Intelligence gathering and social communication in India, 1780-1870.* Cambridge: Cambridge Studies in Indian History and Society, 45.

258 Glaisyer, N. (2004). Networking: Trade and exchange in the eighteenth-century British Empire. *The Historical Journal,* 47, 451-476.

259 Raudzens, G. (1999). Military revolution or maritime evolution? Military superiorities or transportation advantages as main causes of European colonial conquests to 1788. *The Journal of Military History,* 63, 631-641.

260 Raudzens, G. (1999). Military revolution or maritime evolution? Military superiorities or transportation advantages as main causes of European colonial conquests to 1788.

261 Raudzens, G. (1999). Military revolution or maritime evolution? Military superiorities or transportation advantages as main causes of European colonial conquests to 1788.

[262] Raudzens, G. (1999). Military revolution or maritime evolution? Military superiorities or transportation advantages as main causes of European colonial conquests to 1788.

[263] O'Brien, P.K. (2011). The contributions of warfare with revolutionary and Napoleonic France to the consolidation and progress of the British Industrial Revolution. *London School of Economics, Department of Economic History, working papers* no. 150/11.

[264] Van de Ven, J. (1999). The British Industrial Revolution: Joel Mokyr. *NAKE Nieuws, Workshop Report*, 11.

[265] O'Brien, P.K. (2011). The contributions of warfare with revolutionary and Napoleonic France to the consolidation and progress of the British Industrial Revolution.

[266] O'Brien, P.K. (2011). The contributions of warfare with revolutionary and Napoleonic France to the consolidation and progress of the British Industrial Revolution.

[267] O'Brien, P.K. (2011). The contributions of warfare with revolutionary and Napoleonic France to the consolidation and progress of the British Industrial Revolution.

[268] Ang, J.B., et al. (2010) Innovation, technological change and the British Agricultural Revolution. *The Australian National University, Centre for Applied Macroeconomic Analysis*, Working Paper No. 11/2010, 1-27.

[269] Ang, J.B., et al. (2010). Innovation, technological change and the British Agricultural Revolution.

[270] Bettencourt, L.M.A. et al. (2007). Growth, innovation, scaling, and the pace of life in cities. *Proceedings of the National Academy of Sciences*, 104, 7301-7306.

[271] Bettencourt L.M.A., & West, G.B. (2010). A unified theory of urban living. *Nature*, 467, 912-913.

[272] West, G.B. (1999). The origin of universal scaling laws in biology.

[273] Tero, A., et al. (2010). Rules for biologically inspired adaptive network design. *Science*, 327, 439-442.

[274] Tero, A., et al. (2010). Rules for biologically inspired adaptive network design.

Chapter 13

[275] Weinberger, D. (2011). The machine that would predict the fu-ture. *Scientific American*, December, 52-57.

[276] Kurzweil, R. (2005). *The Singularity is near*. New York: Penguin.

[277] Johnson, S. (2010). *Where good ideas come from: The natural history of innovation*. New York: Riverhead Books.

[278] Bear, M.F, Connors, B.W., & Paradiso, M.A. (2001). *Neuroscience: Exploring the brain*. Baltimore: Lippincott.

[279] Roth, G., & Dicke, U. (2005). Evolution of the brain and intelligence trends. *Cognitive Sciences*, 9, 5.

[280] Greg Mayer, "How, and how fast, did the human brain evolve?

[281] Greg Mayer, "How, and how fast, did the human brain evolve?"

[282] Clark, DD; & Sokoloff, L. (1999). Circulation and energy metabolism of the brain. In G. Siegel et al (Ed.s), *Basic Neurochemistry: Molecular, cellular and medical aspects* (pp. 637-670). Philadelphia: Lippincott.

[283] McAuliffe, K. (2011). If modern humans are so smart, why are our brains shrinking?. *Discover Magazine*.

[284] Falk, D., Lepore, F.E., & Noe, A. (2012). The cerebral cortex of Albert Einstein: A description and preliminary analysis of unpublished photographs. *Journal of Neurology*, 1-24.

[285] Leviton, R. (1995). *Brain builders! A lifelong guide to sharper thinking, better memory, and an age-proof mind*. New York: Penguin, 1995.

[286] Rabeling C., Brown J.M., & Verhaagh M. (2008). Newly discovered sister lineage sheds light on early ant evolution. *Proceedings of the National Academy of Sciences*, 105, 1-5.

[287] Sears, C.L. (2005). A dynamic partnership: Celebrating our gut flora. *Anaerobe*, 11, 247-251.

[288] Savage, D.C. (1977). Microbial ecology of the gastrointestinal tract. *Annual Review of Microbiology*, 31, 107-33.

[289] Steinhoff, U. (2005). Who controls the crowd? New findings and old questions about the intestinal microflora. *Immunology Letters*, 99, 12-16.

[290] Berg, R. (1996). The indigenous gastrointestinal microflora. *Trends in Microbiology*, 4, 430-435.

[291] Ndabahaliye, A. (2002). Number of neurons in the human brain. *The Physics Factbook*, http://hypertextbook.com/facts/2002/AniciaNdabahaliye2.shtml.

[292] (1999). Wal-Mart sets the standard for supply chain automation. *Accounting Software Research: Insights into Accounting Systems and ERP Research*, http://www.asaresearch.com/ecommerce/supplychain.htm.

[293] Evans, B. (2007). Wal-Mart's latest 'Orwellian' technology move:

Get over it. *InformationWeek*,
http://www.informationweek.com/wal-marts-latest-
orwelliantechnology-mo/198800804.

[294] Gunther, M. (2007). Wal-Mart sees green. *CNN*.

[295] Souder, E. (2008). Will Wal-Mart sell electricity one day? *Red Orbit*.

[296] Staff Writer (2005). Is Wal-Mart Going Green. *MSNBC*.

[297] (2007) Wal-Mart rolling out new company slogan. *Reuters*.

[298] Talley, K. (2009). UPDATE: Wal-Mart giving US employees $2B in yearly award program. *The Wall Street Journal*,
http://online.wsj.com/article/BT-CO-20090319713086.html?mod=.

[299] Walmart. (2010). *Walmart financial highlights*.
http://walmartstores.com/sites/annualreport/2010/financial_high
lights.aspx

[300] Forbes Magazine. (2010) Forbes the world's billionaires. *Forbes*,
http://www.forbes.com/lists/2010/10/billionaires-2010_TheWorlds-
Billionaires_Rank.html.

[301] Gomstyn, A. (2010). Walmart CEO pay: More in an hour than workers get all year? *ABC News*,
http://abcnews.go.com/Business/walmart-ceo-pay-hourworkers-
year/story?id=11067470#.UVTR7qtUNYg.

[302] Jackson, J. (2011) .IBM Watson vanquishes Human Jeopardy Foes. *PC World*,
http://www.pcworld.com/article/219893/ibm_watson_vanquishe
s_human_jeopardy_foes.html

[303] Mearian, L. (2011). Can anyone afford an IBM Watson super-computer? (Yes). *Computerworld*,
http://www.computerworld.com/s/article/9210381/Can_anyone_
afford_an_IBM_Watson_supercomputer_Yes_?taxonomyId=67pa
geNumber=2.

[304] Thompson, C. (2010). Smarter than you think: What is I.B.M.'s Watson? *The New York Times Magazine*,
https://www.nytimes.com/2010/06/20/magazine/20Computer-
t.html.

[305] Upbin, B. (2011). IBM's supercomputer Watson wins it all with $367 bet. *Forbes*,
http://www.forbes.com/sites/bruceupbin/2011/02/16/watsonwins-it-
all-with-367-bet/.

[306] IBM (2011) Watson - a system designed for answers: The future of workload optimized systems design. *IBM Systems and Technology, An IBM White Paper*.

[307] Gaudin, S. (2011). IBM's Watson could usher in new era of medicine. *Computerworld.*

[308] Krishnaraj, A. (2011). Will Watson replace radiologists? *Diagnostic Imaging.*

[309] Martin, K.L. (2011). Trendspotter: Temporary and tech solutions to the physician shortage. *Physicians Practice.*

[310] Null, G. et al. (2010). *Death by medicine.* Mount Jackson: Pratikos Books.

[311] American Association for Justice (2011). *Medical Negligence Primer.*

[312] Null, G., et al. (2010). *Death by medicine.*

[313] Schrank, D., Lomax, T., & Eisele, B. Texas (2011). *Urban mobility report.* Texas Transportation Institute.

[314] Schrank, D., Lomax, T., & Eisele, B. Texas (2011). *Urban mobility report.* Texas Transportation Institute.

[315] Williams, P.F. (2009). Street smarts: How intelligent transporta-tion systems save money, lives and the environment. *ACS Transportation Solutions Group.*

[316] Monahan, T. (2007). "War rooms" of the street: Surveillance practices in transportation control centers. *The Communication Review*, 10, 367-389.

[317] Johari, J.A.Y., & Al-Khateeb, K. (2008.) Intelligent dynamic traf-fic light sequence algorithm using RFID. *ICCCE*, 1367-1372.

[318] Williams, P.F. (2009). Street smarts: How intelligent transporta-tion systems save money, lives and the environment.

[319] Ezell, S. (2010). *Intelligent transportation systems.* The Information Technology and Innovation Foundation, Explaining International IT Application Leadership.

[320] Williams, P.F. (2009). *Street smarts: How intelligent transportation systems save money, lives and the environment.*

[321] Williams, P.F. (2009). *Street smarts: How intelligent transportation systems save money, lives and the environment.*

[322] Njord, J. et al. (2006). Safety Applications of intelligent transportation systems in Europe and Japan. *International Technology Scanning Program.*

[323] Monahan, T. (2007). "War rooms" of the street: Surveillance practices in transportation control centers

[324] Kurzweil, R. (2005). *The Singularity is near.* London: Penguin Group.

[325] Dooling, R. (2008). *Rapture for the geeks: When AI outsmarts IQ.* Random House Digital.

[326] Kurzweil, R. (2005). *The Singularity is near.*

[327] Kurzweil, R. (2005). *The Singularity is near.*

[328] Gregory, M. (2011). Unemployment is world's fastest-rising fear survey. *BBC News,* http://www.bbc.co.uk/news/business16108437.

[329] Johnson, S. (2010). *Where good ideas come from.*

www.ingramcontent.com/pod-product-compliance
Lightning Source LLC
Chambersburg PA
CBHW070524220526
45467CB00003B/831